COMMUNIST MILITARY MACHINE

COMMUNIST MILITARY MACHINE

General Editor: IAN BECKETT

Bison Books

Published by
Bison Books Ltd.
176 Old Brompton Road,
London, SW 5
England.

ISBN 0 86124 220 3

Printed in Hong Kong

Reprinted 1987

Page 1: A Frelimo soldier on guard at a rail depot near the Mozambique-Zimbabwe border.
Page 2-3: Soviet Tu-95 Bear-D long-range bomber.
This page: Tanks and infantry of the Leningrad Military District during an exercise in February 1985.

Contributors

Dr Ian Beckett is Senior Lecturer in War Studies at the Royal Military Academy, Sandhurst and a Fellow of the Royal Historical Society. He is the author of *Riflemen Form* and co-editor of *Politicians and Defence, Armed Forces and Modern Counter-insurgency*, and *A Nation in Arms.* He also contributed to the recent highly-successful *British Military Operations, 1945-84.* Dr Beckett is currently completing a study of the Curragh Incident of March 1914 for the Army Records Society. (*Introduction, Chapters 6 & 7, Conclusion.*)

Stephen Dalziel has been a student of Russian and Soviet affairs for many years, and took his degree in Russian Studies at the University of Leeds. He has visited the Soviet Union regularly, and has studied in Moscow, Leningrad and Kiev. He has studied also at Sofia University in Bulgaria. Since July 1982, he has been employed as a researcher and lecturer at the Soviet Studies Research Centre at the Royal Military Academy Sandhurst. (*Chapters 1 & 2.*)

Contents

Major F A Godfrey, MC, served in Malaya, Cyprus, Malta, Libya and Aden before retiring from the British Army in 1969. From 1973 to 1982 he was a Senior Lecturer in the Department of War Studies and International Affairs at The Royal Military Academy, Sandhurst. He has contributed to books on the Vietnam and Korean Wars as well as the recent *British Military Operations, 1945-84*. (*Chapters 3 & 4*.)

David Rosser-Owen is a former infantry officer of the British Army. After leaving the regular army, he has worked as a defense journalist specializing in the Islamic world with particular reference to the Middle East and Southeast Asia. (*Chapter 5*.)

INTRODUCTION

Like other mass political movements of the twentieth century, communism is essentially a product of the nineteenth century. What may be broadly described as modern socialism has its origins in the writings of Frenchmen such as Pierre-Joseph Proudhon and Louis Blanc in the 1840s if not in the earlier work by such theorists as Jean-Jacques Rousseau and St Simon. Karl Marx and Friedrich Engels, whose co-written *Communist Manifesto* was published in 1848, were in the mainstream of this development though they were to forge their own particular brand of revolutionary socialism. Their famous slogan, 'Workers of the World, Unite! You have nothing to lose but your chains' echoes Rousseau's 'Man is Born Free, but everywhere he is in chains' in an unconscious ancestry. Nor can Marxism be divorced from the age in which it was conceived – one of unfettered capitalism – or from the intellectual climate of German philosophy as represented by theorists such as Georg Wilhelm Friedrich Hegel. Marx may also have been influenced by his own Jewish background. But Marxism has continued to develop since Marx's day in the sense that the original orthodoxy has been constantly refined and redefined by others such as Vladimir Ilyich Ulyanov, better known as Lenin, Mao Tse-tung, Fidel Castro and the North Korean leader, Kim Il Sung.

In its original form, Marxism professed to be a scientific explanation of the course of history and the development of society, providing not just a political ideology but a system of government. Reasoning that the determinant of all human action was the economic self-interest of various social groups or classes, and that a 'ruling class' would so order the economic system as to control and exploit the other classes, Marx evolved an analysis of the course of history known as the dialectic. In the dialectic, society moved inexorably and inevitably from primitivity through slavery and feudalism to capitalism and, ultimately, through a transitional stage of socialism to true communism. Marx believed that society in his age was at the stage of capitalism, in which economic exploitation had led to class struggle and which, through its inherent 'contradictions,' was doomed by the force of history. However, Marx did not believe that the transition from capitalism to socialism, in which private ownership of the means of production and labor would be no longer tolerated and in which rewards would be proportional to the value of the labor performed, could be peacefully attained. Therefore, the only way forward would be by revolution to wrest control from the ruling class and its bourgeois allies, conceivably through the

union of the exploited or 'proletarian' classes of the world. After the revolution, which would only occur at an appropriate moment, there would be the transitional stage of socialism. The apparatus of the state would survive during the period of socialism and the state would be ruled by the proletariat but, in the final stage of communism, all peoples would live free of the shackles of the state and undivided by any class interests.

Marxist theory was at times somewhat vague, Lenin's contribution being primarily the practical means by which revolution might actually be achieved. That means would be through the actions of the organizational weapon of a Communist Party acting as a 'vanguard' of the proletariat and, in the stage of socialism, becoming the instrument of the 'dictatorship of the proletariat,' by which opponents of the new order would be defeated and society transformed. In Marxist-Leninist terms, the modern Soviet Union is regarded as being in transition from socialism to communism while the remainder of the world is in transition from capitalism to socialism. The existence of the 'socialist bloc' led by the Soviet Union, together with the working classes of existing capitalist countries and the movements for national liberation in the Third World, mean that communism will ultimately triumph. There may be temporary setbacks in the short-term, because capitalism is strong and has the capability of ensuring mass destruction through nuclear war, but the final triumph is assured.

Above: Vladimir Ilyich Lenin (1870-1924) as depicted on the badge of the quasi-military Soviet youth organization, the Pioneers.

Left: Another Soviet poster from World War II invoking Russian nationalism by showing the eighteenth-century Tsarist general, Suvorov, encouraging the Red Army.

Far left: A propaganda poster from the 'Great Patriotic War' (1941-1945), which marked the real emergence of the Soviet Union as a global power.

Above: Friedrich Engels (1820-1895), whose co-operation with Marx resulted from his acting as Manchester agent for his father's textile business.

Communism is, however, nothing if not adaptable. Joseph Stalin evolved the concept of 'socialism in one country,' since it was apparent that the theory of 'permanent revolution' as advocated by Leon Trotsky and others no longer 'fitted the facts' of the Soviet Union's isolation in the 1920s. Lenin, too, had accommodated the European imperialism of the late nineteenth century, which Marx did not witness, within the context of Marx's economic framework of historical analysis. At the same time, Stalin did not abandon the concept of inevitable conflict with capitalism and his idea of what would later resemble Nikita Khrushchev's 'peaceful co-existence' with the West was juxtaposed with dogmatic assertions about the likelihood of war. In other countries, too, Marxist-Leninist theories have been adapted to fit local circumstances, most notably in China.

Ideology lies at the heart of the Sino-Soviet rift, which emerged publicly in 1960, although policy differences were also apparent. Mao Tse-tung, for example, was hostile to Khrushchev's 'peaceful co-existence' in the 1950s and the Soviets were generally more cautious with regard to the promotion of revolution in the emerging Third World than Mao. Maoist thought, of course, was also derived from the reliance necessarily placed upon the peasantry in China during the course of the Chinese revolution since Leninist models of urban insurrection by an urban industrial proletariat were simply not appropriate. Maoism similarly embraced a distinctly moral attitude, partly derived from ancient Chinese traditions, which was alien to the Marxist 'scientific' approach, Maoism stressing the ability of the individual to escape the beliefs inherited from class background through what is known as a 'redemptive historical process.'

To the Chinese communists, the world is divided into three distinct zones. The first, to which only China herself is qualified to belong, is the 'socialist camp.' At one time Albania was also accorded this distinction but has since been deemed to have become revisionist. Similarly, a number of suitably progressive communist parties such as that of New Zealand, are also counted as part of the socialist camp, which is characterized as progressive, revolutionary and a proponent of peace and freedom in which all forms of class and colonial exploitation have been eliminated. The second zone is the 'imperialist camp,' which consists of the two superpowers of the United States and the Soviet Union and their satellites and allies. Aggressive and greedy, the imperialist powers can only sustain themselves through domination of smaller states. They are in perpetual conflict with each other but, since they have some common interests, will also protect these mutual interests by conspiring to prevent the emergence of a stronger socialist camp. Such co-operation represents a 'world condominium conspiracy.' The third area is the 'intermediate zone,' potential victim of the exploitation of the imperialists, which consists of the Third World and western Europe. The Chinese believe that they must promote solidarity in the Third World and encourage the Europeans to build up their strength so that both may be safe from the designs of the imperialists.

Nevertheless, as the recognition of capitalist western Europe as a potential friend and the rapprochement with the United States since 1971 indicates, the Chinese have developed considerable pragmatism in recent years. They still believe, however, that there must ultimately be revolution in both the United States and the Soviet Union if a world war is to be avoided. The Chinese do not fear such a war since 'fear solves no problem whatsoever' but they also believe that it can be postponed by defensive preparations, which would enable states to maintain their independence from the imperialists. In their own analysis, the Chinese have thus consistently used military power defensively since the establishment of the People's Republic in 1949. They would argue that the often ruthless suppression of non-Chinese peoples in frontier areas such as Manchuria, Mongolia, Sinkiang and Tibet maintains the security and territorial integrity of China. Intervention in the Korean War in 1950 was regarded as necessary in defense of China itself against the aggressive intentions of the United States, in the guise of the United Nations forces, driving toward the Yalu. Since the Chinese regard themselves as the legal heir to Imperial China, the occupation of 'lost' territories like Tibet and Sinkiang in 1950 and the brief border war with India in 1962 merely secured traditional frontiers. The Sino-Soviet frontier is one factor in the Sino-Soviet split while the Chinese invasion of Vietnam in 1979 also had territorial implications as well as being a response to a threat posed by a client of the Soviet Union. In the 1950s and 1960s, the Chinese sponsored 'wars of national liberation' throughout Southeast Asia and in other parts of the world but these were regarded as a legitimate means of removing threatening alien influences from Asia. Overt support for such groups in Southeast Asia has now been abandoned as has the intention to invade Taiwan and reconquer it by force of arms, this being part of the rapprochement with the United States.

The Soviet Union, too, has shown a preference for pragmatism and compromise with the West on occasions as evinced by Khrushchev's peaceful co-existence and the process of 'detente' associated with the Brezhnev years. However, it should not be forgotten that in September 1968, following the Soviet invasion of Czechoslovakia, the Kremlin annunciated what has become known as the 'Brezhnev doctrine,' also referred to as the 'doctrine of the socialist commonwealth' and the 'doctrine of limited sovereignty.' By this the Soviets reserved the right to intervene whenever and wherever it was deemed that socialism or the interests of other socialist states were threatened by internal developments in any one country. In theory this contradicted the preamble to the Warsaw Pact treaty of 1955 and was dismissed by Leonid Brezhnev in 1971 as a 'Western fabrication' but it is still very much alive. The precise limits of the 'socialist commonwealth' have never been defined. It clearly includes not only eastern Europe but Afghanistan and, presumably, Albania and Yugoslavia or any state in which communism has achieved power or may achieve power in the future. 'Fraternal' assistance has, of course, been extended to a number of states such as

Above: Karl Marx (1818-1883), pictured during his exile in England from 1849 onward, much of his time being spent in the reading room of the British Museum.

Left: Josef Visarionovitch Stalin (1879-1953), the Georgian-born successor to Lenin as Soviet leader, photographed with Orjouikidze and Voroshilov at the 1st Conference of Stakhanovites in November 1935.

НАРОД И ПАРТИЯ ЕДИНЫ!
СЛАВА КПСС!
XXVI СЪЕЗДУ КПСС-ДОСТОЙНУЮ ВСТРЕЧУ!

Above: A contemporary poster celebrating the armed forces that contribute so much to the strength of the Soviet Union.

Left: The military parade in Moscow's Red Square held on 7 November annually to mark the anniversary of the Russian Revolution.

Afghanistan, Angola and Ethiopia and the treaties of friendship and co-operation concluded with these and other states provide for such a contingency. The Soviets have consistently refused to abrogate a 1921 treaty with Iran which allows for their intervention there in certain circumstances.

Article 28 of the new Soviet constitution, adopted in 1977, is also significant. Unusually for a constitution, this commits the Soviet government to precise foreign policy aims:

> The foreign policy of the USSR is aimed at ensuring international conditions favorable for building Communism in the USSR, safeguarding the State interests of the Soviet Union, consolidating the position of world socialism, supporting the struggle of people for national liberation and social progress, preventing wars of aggression, achieving complete and universal disarmament, and consistently implementing the principle of peaceful co-existence of States with different social systems.

The emphasis lies not in the phrases relating to co-existence and universal disarmament but in the stated goals of building communism and safeguarding the interests of the Soviet Union by the consolidation of world socialism. The Soviets, in pursuit of such goals, do not believe in the 'balance of power' but in the 'correlation of forces,' in which opportunism plays an integral role. Moreover the use of military power has never been renounced by any Marxist. War is regarded as evil and essentially as a product of capitalism but it is also seen by the Soviets as a legitimate tool of policy. Indeed, in a Clausewitzian way, it is a continuation of policy just as peace or neutrality might equally be pursued as useful tools of policy in different circumstances. The process of détente pursued by Brezhnev in the years between 1969 and 1979, in his own words, 'does not in the slightest abolish, nor can it abolish or change the laws of class struggle.' The Soviet Union is thus not embarrassed by military power and the demonstration of that power as it showed in East Berlin in 1953, in Hungary in 1956, in Czechoslovakia in 1968 and in Afghanistan in 1979. Military power will not be used unless victory can be reasonably guaranteed in advance but it remains both a tool of Soviet foreign policy and an agent of the global revolutionary process.

There has often been a lively debate in the West on the extent to which ideological considerations still guide Soviet or, for that matter, communist policies generally. Pragmatism and purely national interests are always bound to play an important part in the calculations of any state's foreign policy and in its reliance on military power to achieve those aims. But the greatest mistake that can be made is to assume that, because of such factors, Marxists do not mean what they say about their ultimate goals. They do.

Ian Beckett

1. THE SOVIET UNION

Introduction

For centuries Russians have lived with a fear that foreign powers will attack and attempt to overrun their Motherland, and, as a result, they have developed a keen awareness of the need to be able to defend themselves. This feeling is not simply paranoia. For 250 years in the Middle Ages they lived under the yoke of Mongol rule, the strictness of which set the pattern for all Russian and Soviet rulers since. Until Peter the Great finally achieved victory over Sweden in the eighteenth century, the north of Russia was under constant threat from that country. In 1812, in what the Russians call the first Great Patriotic War, Napoleon reached Moscow and burned the Kremlin; the French were fighting on Russian soil again forty years later, when, with the British, they fought the Russians in the Crimea. The eastern part of this huge country was not safe either, as the Japanese proved in 1905 after winning the Russo-Japanese War, when they seized territory the Russians regarded as their own. Nine years later the Russian Army was engaged in combat once again, this time against the Germans. The advent of Soviet power in 1917 saw an intensifying of the Russians' fear of attack from

outside, because to the old fears was added an important new political one: an inherent part of the socialist ideology which the Bolsheviks professed, was that the capitalist states were totally hostile to the socialist ideal, and therefore war between capitalism and socialism was inevitable.

According to the ideology of communism, mankind's march toward the communist ideal is unavoidable, and certain steps must necessarily be followed. The crucial stage is the transition from capitalism to socialism. It was believed by the Bolsheviks in 1917 that it would be impossible for this to be achieved by peaceful means, since those who were about to lose power, the capitalists, would fight to the death to hold on to it; and, on the other side, the advocates of socialism would be equally prepared to fight to the death, because they had nothing to lose and everything to gain. Furthermore, they held that this showdown between capitalism and socialism must happen all over the world more or less simultaneously. When this did not occur, the Russian socialists had two courses open to them: to admit the failure of the Revolution, establish some form of capitalist government in the newly-proclaimed Soviet Union, and sit back and wait for the time to be

Opposite page: November 1917, Palace Square, Petrograd. Bolsheviks address a crowd of soldiers and workers after their seizure of power. Under the banner copies of their newspaper *Pravda* (Truth) are for sale.

Below: November 1917, recruits for the Bolshevik Red Guard gathered in a Petrograd railway station.

ripe for the world-wide revolution; or admit that unforeseen factors had delayed the spread of the Revolution, but, nevertheless, it must be consolidated where it had started. Thus was born the doctrine of 'socialism in one country.' However, inherent in the new doctrine was the idea that this was only a temporary state of affairs, and that sooner or later the time would come when the Revolution must be spread, and thus war between capitalism and socialism was still inevitable. In fact, the existence of but one socialist state was thought to make it even more likely, as the capitalists would wish to try to prevent the spread of socialism by snuffing it out at its source: to a Soviet communist, foreign intervention by the British, French, American and Japanese armies in the Civil War of 1918-20, and Hitler's invasion of the USSR in 1941 prove just that.

If you are a member of the Politburo looking at the world map today from Moscow, it looks very different from the way it looks from London, Washington, or anywhere else in the western world. To the west, beyond your Warsaw Pact allies, you see capitalist European countries, bound together militarily by NATO, which is overtly designed to combat the socialist countries of Eastern Europe. NATO's influence extends to Turkey, on the southwestern border of the USSR, alongside which you see Iran, a country governed by a fanatical Moslem regime, and aggressively anti-Soviet. Due south of the USSR is the Democratic Republic of Afghanistan, whose revolution the imperialists are trying to destroy; if they did they would gain a further valuable foothold from where they could strike at the Soviet Union. In the southeast you see China, a country with a population almost four times that of the USSR. This overcrowded nation, you think, may well feel the need to expand her territory to house this population before too long, and how else can she expand but into Soviet territory? Furthermore, she claims to be ruled according to the decrees of Marxism-Leninism, and says that she, and not the Soviet Union, is true to these teachings, thus presenting herself as a rival for the position of world leader of communism. Though a small country, Japan to your east has shown already three times this century that she is prepared to attack the USSR, and her strong position in the capitalist world is enough in your eyes to imply that the threat still exists. And then to the northeast and across the top of the globe stands the greatest threat of all to the security of the USSR and the ideology of Marxism-Leninism – the USA. Thus, with all these potential enemies surrounding you, is it so surprising that you feel a need for large armed forces in order to be able to protect yourself?

Be it for traditional Russian reasons or modern Soviet ones, therefore, the Soviet Armed Forces have a vital role to play in the USSR today. However, just as it is crucial to examine the *raison d'être* of the Soviet Armed Forces from a Soviet point of view in order to understand their defensive role vis-à-vis the USSR's own borders, it is also vital to look at

Above right: Workers' demonstration in Petrograd, 1917. The banners proclaim 'Down with Ministers and Capitalists! All Power to the Soviets!.'

Below right: Demonstrators flee the streets of Petrograd as machine gunners open fire.

their political role from this standpoint, too. When the task of the Soviet Armed Forces is described as 'the defense of the Revolution,' it must be remembered that communism is an internationalist philosophy, which claims to override national boundaries. Therefore, if the Soviet Union is called upon to help defend the Revolution in Afghanistan, the arrival of Soviet troops in that country is not seen by the Soviet leadership as being 'foreign intervention'; it is merely an extension of the main principle upon which the Soviet Armed Forces lie: the defense of the Revolution.

It is this political factor which is the marked difference between the armed forces of a communist state and those of a western one. In a western state the army must be apolitical, as it must obey the commands of whichever political party happens to be in power. In a Marxist state, there can only ever be one party in power, the Communist Party, and so the army becomes a political one. Nevertheless, it is important to remember that it must always be the Party which has the leading role, and, however strong the army may be, it is always governed by the Party. This is due to Lenin's adoption of Clausewitz's dictum that 'war is the continuation of policy by other means'; thus, the military machine will roll into action only as and when the Party requires it to do so, so as to attain an objective of policy.

History

The birth of the Red Army provides the classic example of revolutionary idealism giving way to practical reality. While revolutionaries may dream of overturning society and creating a new one in its place, time does not stand still to enable them to carry this out. Thus, as life must go on before, during, and after the revolution, the only way to maintain any order is to adapt the organs of the old state, rather than to destroy them totally. The Bolshevik leaders quickly realized that this was the case with their 'new' army.

Since the aim of the Great October Socialist Revolution was to create a republic of workers and peasants, it was naturally assumed that the Red Army would be an army composed of members of those classes. The seeds of this class conscious force were sown after the February Revolution of 1917 (which overthrew the Tsar), when the Bolsheviks created volunteer armed detachments at factories and in Party committees throughout the country, under the title of Red Guards. Almost every city in the country had one or more detachments of Red Guards by the time of the October Revolution. At the same time, Bolsheviks serving in the Russian Army and Navy were working hard to recruit soldiers and sailors to their cause. It was thought, therefore, that when the Bolsheviks seized power the enthusiasm of these workers and peasants would be enough to protect and spread the Revolution throughout the country. The new Red Army refused to accept anyone who came from the 'reactionary' sections of pre-Revolutionary Russian society. As a result, by the end of 1917 thousands of officers of the Imperial Russian Army and Navy were removed and replaced by Red Guards. The decision as to who was to serve as an officer was made by a vote of the men. On 29 December 1917, the ranks of the old army were abolished. However, by early 1918, the Bolshevik military leaders were beginning to realize that the honeymoon of the Revolution was over, and that if they were to survive they would have to have a competent army under the guidance of professional military men.

The Workers' and Peasants' Red Army was officially formed on 28 January 1918, and following this there was mass support for Lenin's call to defend the socialist Motherland, which led to 23 February being declared as the day of the Defense of the Motherland (now Soviet Armed Forces Day). Despite this, the Bolsheviks understood that without competent military leadership the enthusiasm of their soldiers would be crushed by the German Army from without, and counter-revolutionary forces from within Russia. On 3 March 1918, in an attempt to stave off the external threat, the Soviet leadership ceded vast areas of land to the Germans when they signed the Treaty of Brest-Litovsk, which took Russia out of World War I. Later that month the recruitment of former Imperial Army officers began, which underlined the need to bow to reality. About 22,000

joined. Nevertheless, it was believed that these officers could pose a threat as potential saboteurs, and so at the same time the system of political commissars was introduced. The commissars' principal role was to ensure the loyalty of the former Tsarist officers, but, given the totally political nature of the Red Army, they had the tasks also of initiating political education classes for the soldiers and of enforcing discipline. The importance of the commissar was emphasized not only by his position outside the jurisdiction of the unit commander, but also by the requirement that, at regimental and divisional levels, all orders to subordinate personnel had to have his assent and signature.

The spring of 1918 was a crucial period for the development of the Red Army, under the guidance of its leader Trotsky, officially appointed Chairman of the Higher Military Council in March. Although the immediate external threat to the state had apparently been ended at Brest-Litovsk, foreign intervention began in April, in places as far apart as Archangel and Murmansk in the north, eastern Siberia, Baku on the Caspian Sea, and the Black Sea ports of Batumi, Odessa and Sevastopol. Although the intervention proved to be half-hearted and uncoordinated, it posed a menace to the infant Soviet state at the time, and is still regarded as an example of the permanent threat of invasion under which Russia has lived for centuries. By May 1918 the Red Army numbered some 300,000 volunteers, which clearly was not enough to hold off not only foreign invaders, but also 'the Whites,' former Imperial Army soldiers and counter-revolutionaries who were to continue the Civil War for more than two and one half years. Compulsory military service for all males between the ages of 18 and 40 was introduced, and discipline was tightened. This included making virtual hostages of the former Tsarist officers, as they were put under threat of treason or desertion meaning the execution of their families.

The sheer size of Russia and the extent of the fighting in the Civil War meant that it was vital for the Soviet leadership to establish a strict form of coordination and control for the Red Army. Republican Revolutionary Military Soviets were set up in all fronts and armies, and the Red Army was placed under the overall command of Vatsetis, a former colonel in the Tsarist Army. After he had collected together a field staff which included other ex-Imperial officers, for the coordination of all the armies and fronts, Trotsky realized just how essential it was to have competent 'military specialists' as commanders, and so dismissed most of the self-appointed Bolshevik commanders, and put in more professional military men who had served in the Tsar's Army. Nevertheless, they were still watched carefully by the commissars.

The success of the reorganization was proved by the Bolsheviks' achievement in turning a situation where in late 1918 three-quarters of the old Russian Empire was in the hands of the Whites or under foreign occupation, to one

Above: Red Army recruits being trained in the use of a rifle. The painting on the wall differentiates between the straight line of aim, and the curved line of the trajectory of the bullet, which has a range of 560 paces.

Opposite page, top: Part of the huge funeral procession which followed Lenin's coffin after the death of the leader of the Revolution on 20 January 1924.

Opposite page, bottom: Early Red Army infantry, armed with the M-1930 7.62mm rifle.

Left: Bolshevik revolutionaries open fire on the police in the early days of the Revolution.

where they had achieved complete victory just two years later. Organization and coordination of the military effort was one of the Reds' strengths; lack of it was one of the Whites' greatest weaknesses. By the time the fighting ended in October 1920, the Red Army stood at nearly five and one half million strong. Not all of these were engaged in combat, however. From March many thousands had been engaged in helping to rebuild the shattered civilian economy, thus setting a precedent for the Soviet conscript today, who, as well as his normal soldiering, may well be called upon to gather the harvest, repair roads, or to assist in the construction of buildings or other economic projects, such as the construction of the new Baikal-Amur Railroad.

With the Civil War over, mass demobilization of the Red Army began. One of the first actions, in January 1921, was the transfer of some of the best units to the secret police (then called the Cheka) to act as border guards. This procedure has continued to the present: while carefully chosen because of their military and political prowess, the Soviet Border Guard Troops are not part of the Ministry of Defense, but of the KGB – the Committee for State Security. Demobilization continued to the point where, by October 1924, the Red Army was down to 545,000 men. That the numbers were allowed to be reduced so much was due to the need to restore the civilian economy, not because it was felt that large armed forces were unnecessary. On the contrary, because Marxism-Leninism predicted the inevitable war between capitalism and socialism, the armed forces needed to be made as strong as possible as quickly as possible. This did not simply mean numbers of men, however. The Red Army was in need of equipment and military training and education for its officers and men. Of the European powers, only Germany was prepared to provide such help to a socialist state, because Germany was not only in the position also of

This page: Soviet Border Guards defending the frontiers of the Soviet Union. These troops, easily distinguishable by their green epaulettes, are the first that the visitor to the USSR sees; yet they do not come under the command of the Ministry of Defense, but the Committee for State Security (KGB).

Opposite page, top: Because the Soviet Army is tasked to fulfil the wishes of the Communist Party of the Soviet Union, political education of troops has always been accorded an important role in the training of the Soviet soldier. This takes place not only in the classroom, but in the field as well. Here a political worker NCO reads the day's *Pravda* to soldiers. In the background he has a stand showing achievements of the Soviet Union.

Opposite page, bottom: Soviet infantry on exercise. The soldier in the foreground is using the AKM assault rifle designed by M. T. Kalashnikov. Light and easy to handle, this weapon is used by forces the world over, and is sometimes known as 'the terrorist's favorite weapon.'

being an outcast after World War I, but was forbidden by the Treaty of Versailles to carry out military production. An agreement with the Russians to cooperate on military development was seen as mutually beneficial on purely practical grounds. Thus it was that the Treaty of Rapallo was signed in 1922. This led to the construction of a Junkers aircraft plant in Moscow, and the setting up of a flying school for German pilots near Lipetsk (southeast of Moscow), and tank and chemical warfare training establishments. Russian officers attended German staff courses, while German officers took part in field exercises in the Soviet Union.

By the mid-1920s, the Red Army was manned largely on a territorial principle, by administrative region. A man's conscripted service would be spread over five years: the first period would be one of three months, then there would be one month per year to serve for the next four years. To maintain continuity, 16 to 20 percent of each territorial division were regulars. Technical troops, the Navy, and the Border Guards, however, were manned by regular officers and conscripts serving a full two year term, a system which was to be adopted throughout by the late 1930s. Between 1925 and 1930, the commissar system was revised, too. Instead of the political commissar being an overseer of the commander, he was now subordinated to him, and became the deputy commander for political affairs ('*zampolit*'). Although there were brief periods between 1937 and 1942 when the old system was re-adopted, since then one-man command with a *zampolit* has been enforced.

If the twenties was the decade of the consolidation of the civilian economy in the USSR, the thirties saw far greater concentration on the military side. By then the defense industries were the first in line for raw materials (which remains the case today) – production per annum of aircraft, artillery pieces and tanks all increased by approximately four times between 1930 and 1937 – and the numbers in the armed forces grew to over one and one half million by 1938. The Air Forces expanded sufficiently to warrant their division into three branches: bomber, dive-bomber and fighter, and the Navy not only recovered from its post-Civil War position of having virtually every ship damaged, but acquired sufficient new vessels to form two new fleets, the Northern Fleet and the Pacific Ocean Fleet.

Yet another of the measures taken in the immediate aftermath of the Revolution – the abolition of military ranks – was rescinded when Stalin restored all ranks to the Army, beginning in September 1935 (ranks began to be restored to the Navy in December 1936). Furthermore, he named the first five Marshals of the Soviet Union: Voroshilov, Budenny, Tukhachevsky, Blyukher, and Yegorov. The last three were to fall victim to the greatest disaster to strike the

Above: The political nature of the Soviet Armed Forces is illustrated by the large number of Communist Party members among the officers, and Young Communist League (Komsomol) members among the conscript soldiers. Here, at the All-Army Conference of Secretaries of Komsomol Organizations, held in the Kremlin Palace of Congresses in May 1984, the then President Chernenko presents the Komsomol Organization with the banner of the Order of the Soviet Army and Navy.

senior commanders of the Soviet Armed Forces: Stalin's military purges of 1937-38. These were the culmination of the Great Purge which began in 1934, and which enabled Stalin to weed out and eliminate anyone who opposed him in any way, or presented any kind of threat to him, be it real or, as in many cases, imaginary. As well as 3 of his 5 Marshals, Stalin purged 14 out of 16 Army Commanders; all 8 Admirals; 60 out of 67 Corps Commanders; 136 out of 199 Divisional Commanders; 221 of the 397 Brigade Commanders; all 11 of the Deputy Commissars of Defense; and 75 of the 80 members of the Supreme Military Soviet. In all, it is estimated that some 35,000 officers suffered. The political administration of the Army was hit heavily, too. The purge accounted for all 17 Army Commissars; 25 out of the 28 Corps Commissars; and 34 of the 36 Brigade Commissars. Approximately two thirds of the political administration was purged, in total about 20,000 men.

Although the Soviet Armed Forces were undoubtedly weakened by the purges, their effect was not as catastrophic as might have been expected, as the Soviet Army was soon forced to prove. First, in July 1938, fighting broke out between Soviet troops and the Japanese around Lake Khasan. Under the guidance of the then Soviet commander in the Far East, Marshal Blyukher, the Soviet troops forced the Japanese to withdraw. However, a year later there was a further clash, this time at Khalkin Gol, but again the Japanese were forced back. As victory was being achieved, the Nazi-Soviet Non-Aggression Pact was being signed, on 23 August 1939. Clearly neither side expected it to last for its full 10 years, but it gave Stalin a chance to prepare for war. Although the Soviet Union was far from ready when Germany attacked on 22 June 1941, certain measures had been taken. The Soviet Union's European borders had been pushed out by the annexation of the western parts of Belorussia and the Baltic States in 1939 and 1940, and after the war with Finland in the winter of 1939-40 Leningrad was thought to be more secure. Ironically, the poor showing of the Red Army in the Winter War helped to persuade the Germans that an attack on the Soviet Union could bring victory. The day that Germany attacked Poland, 1 September 1939, a new law on universal military service was passed in the USSR, which extended the conscription period from two to five years. By 1941 the Soviet Armed Forces totalled some 4,200,000 men.

The Great Patriotic War of 1941-45 provided the Soviet Armed Forces with their greatest moments of glory, not only in the defense of their own Motherland, but also in the achievement in spreading the Revolution by 'liberating' the countries of Eastern Europe in 1944 and 1945. It is obvious that the Soviet Union wished to form a buffer zone between herself and her potential capitalist enemies in the west, to ensure that there would be no repetition of the carnage wreaked by the Germans when 20,000,000 Soviet citizens perished, and over 70,000 towns and villages, and 32,000 industrial enterprises were destroyed. However, the arrival of the Red Army in Eastern Europe in 1944 also meant that for the first time since 1917 it could carry out what it saw as its historic mission to help the inevitable spread of communism. Without detracting in any way from the suffering undergone by the Soviet Union during the Great Patriotic War, nor the Red Army's remarkable achievement in recovering from an appalling start, and, having stared defeat in the face, in achieving total victory on the Eastern Front, the long term significance of the Great Patriotic War has undoubtedly been the manner in which it so totally changed the map of Europe. Soviet writers make it perfectly clear that, had the Red Army been allowed to reach the English Channel in 1945, it would gladly have done so, because of the ideology which inspires the Soviet leadership. Thus, 1945 is the most significant year in the history of the Red Army since its formation in 1918: not only did it defend the Revolution at home, but it took the first step toward its historical mission of spreading it world-wide, too.

Since 1945 the principal factor in the development of the Soviet Armed Forces has been the coming of the nuclear age, coupled with the establishment of the USSR alongside the USA as one of the world's two superpowers, divided by totally opposing creeds. The sickly Bolshevik baby which had to fight for its very existence some 65 years ago has grown into a giant on the world stage, and it is only by means of strong armed forces that it can guarantee to keep its place there. The post-war development of the Soviet Armed Forces falls into three definable periods: from 1945 until March 1953; from April 1953 to the end of 1959; and from 1960 onward.

The first post-war period (1945-53) was determined by Stalin's personal rule. It was obvious to the Soviet leadership that their wartime ally the United States could, and probably would, become their greatest potential enemy after the war, as she had emerged as the world's most powerful exponent of capitalism. Therefore, the possession by the USA of the atomic weapon meant that it was imperative for the Soviet Union to possess it too. This the USSR achieved by 1949, and yet remained in something of a dilemma: so great was Stalin's control of the whole system in the USSR that, because of his refusal to integrate this new technology into the Soviet military system, the Soviet military leadership had to behave as if it did not exist. This situation changed drastically, however, after Stalin's death in March 1953, which marks the end of the first post-war period and the start of the second. In August of that year the Soviets dropped their first hydrogen bomb, and brought two jet bombers – the Tu-16 (NATO codename Badger) and the M-4 (Bison) – into service. While proceeding with the production of these aircraft, the Soviet nuclear position took another major step forward in 1957, with their first test of an Intercontinental Ballistic Missile (ICBM) in August, and the launch of the world's first artificial satellite on 4 October. This leap in missile technology which meant that US territory was now vulnerable from the Soviet Union significantly and irrevocably altered the strategic balance, and led in the USSR to a greater concentration on missile production than on any other field.

The direct result of this was the 'Revolution in Military Affairs' which was heralded by Khrushchev in January 1960, and which marks the start of the third post-war period. As Soviet military doctrine now held that the next war would begin inevitably with a nuclear strike, and that the nuclear exchange would be the crucial, deciding factor, Khrushchev stressed the need for nuclear weapons to the detriment of all else, manpower included. Although this emphasis, unpopular with the military, was somewhat changed after Khrushchev's removal from power in 1964, the Soviet Union continued to emphasize the 'all-out nuclear' approach until NATO's adoption of the policy of 'flexible response' in 1967, since when Soviet military training has been for the possibility of non-nuclear as well as nuclear war.

In order to meet the apparent threat posed to her security by the United States, the Soviet Union has built up large, nuclear-capable armed forces. However, even if one accepts the idea that these forces have been built up simply for defensive reasons, it is easy to see, both in theory and practice, how useful these armed forces can be to continue the 'historical process' of spreading the Revolution. Since the USSR regards wars of national liberation against capitalist oppressors as just wars, the use of Soviet troops to help the cause is deemed acceptable. However, the geographical and logistical problems involved have meant that there have been few occasions on which Soviet troops have been used outside the country's borders in large numbers. But instances of aid being given by Soviet military advisers, 'friendly visits' by Soviet warships (the post-war development of the Soviet Navy has been remarkable, transforming it from virtually a coastal defense force into the world's second-largest navy), and the encouragement of actual intervention by surrogate troops, such as Cubans, all show that the political role of the Soviet Armed Forces as the defender of the Revolution world-wide has not only remained constant since 1917, but is a far more practical proposition now than at any time.

Overleaf: A banner depicting Lenin, Engels and Marx looks down on a May Day parade on Red Square in Moscow. Like the parade on 7 November, in honor of the Bolshevik Revolution, these parades are designed to glorify the Soviet Union and its achievements. In this parade, one group of marchers form the shape of a rocket, whilst those in the middle make up the word *Soyuz*, the principal designation of Soviet space rockets.

ЗА ЕДИНСТВО
И
СПЛОЧЕННОСТЬ
ВСЕХ
ДА ЗДРАВСТВУЕТ
СОДРУЖЕСТВО
СТРАН СОЦИАЛИЗМА

The Soviet Armed Forces Today

Below: December 1982. Konstantin Chernenko delivers a speech to a combined audience of the Central Committee of the CPSU and the Supreme Soviet of the USSR, at a meeting in honor of the 60th anniversary of the USSR. On assuming the position of General Secretary of the Party in February 1984, Chernenko, like his predecessor Yuri Andropov (seen here sitting on Chernenko's right) automatically became C-in-C of the Soviet Armed Forces too.

Opposite page: Members of the Soviet High Command: (Top left) Marshal of the Soviet Union S. F. Akhromeev, Chief of the General Staff; (Top right) Marshal of Air Forces A. I. Koldunov, C-in-C Air Defense Forces; (Bottom left) Marshal of the Soviet Union S. L. Sokolov, Minister of Defense of the USSR; (Bottom right) Army General A. A. Yepishev, Chief of the Main Political Directorate of the Soviet Army and Navy since 1962.

Organization of the Soviet High Command

All authority in the Soviet Union is organized strictly in pyramid-fashion, and so it is no surprise that this is the case in the Armed Forces, too. The supreme body in determining Soviet military policy is the *Defense Council*, the chairman of which is normally the General Secretary of the Party (currently Mikhail Gorbachev), who thus militarily as well as politically sits at the top of the tree. Although the exact membership of the Defense Council is unknown, it does include other Politburo members who represent the military and economic spheres, as well as the purely political one. Its role is to oversee all aspects of national security policy, including ensuring that standby plans for all areas of society in the event of war are kept up to date.

The primary administration body of the SAF is the *Ministry of Defense*, which is headed in time of peace by the *Main Military Council* (this would be replaced in wartime by the *Stavka*, the Headquarters of the Supreme High Command). Although chaired by the Minister of Defense, the chairman of the Defense Council also has a seat on this body, as do the first deputy ministers of defense – Marshal Akhromeev, Chief of the General Staff (CGS), Marshal Kulikov, Commander-in-Chief (C-in-C) Warsaw Pact Forces, and Marshal Petrov, responsible for general affairs, the Chief of the Main Political Directorate, Marshal Yepishev, and the 10 deputy ministers of defense. These are the C-in-Cs of the five branches of the SAF (the Strategic Rocket Forces, the Ground Forces, the Air Defense Forces, the Air Forces and the Navy), as well as the head of armaments, the inspector general, and the chiefs of civil defense, military construction, and the rear services. Just as the chairman of the Defense Council and the Minister of Defense provide a link between the two highest organs of the SAF, so the CGS and the C-in-Cs of the five services provide the link between the Ministry of Defense and the largest organizational body, the *General Staff*.

A principal reason why membership of the Defense Council and the Main Military Council can be kept to limited numbers is the existence of the General Staff, which deals with operational control over the Armed Forces and determines the everyday roles and tasks of each branch of service. Broadly speaking, the work of the General Staff is primarily in three areas: intelligence, operations, and organization and mobilization of manpower, although there are

Overleaf: Uniforms of the Soviet Army, showing badges of rank and arm of service insignia: (Top left) Sergeants, Conscript Soldiers, and Cadets of the Suvorov Military Colleges; (Bottom left) Praporshchiks (Warrant Officers) and Women's Uniforms; (Top right) Officers (up to Colonel); (Bottom right) Marshals and Generals.

ФОРМА ОДЕЖДЫ И ЗНАКИ РАЗЛИЧИЯ
СЕРЖАНТОВ И СОЛДАТ СРОЧНОЙ СЛУЖБЫ СОВЕТСКОЙ АРМИИ, КУРСАНТОВ
ВОЕННЫХ УЧИЛИЩ, СУВОРОВЦЕВ И ВОЕННЫХ СТРОИТЕЛЕЙ
ПАРАДНАЯ И ПАРАДНО-ВЫХОДНАЯ
ПОВСЕДНЕВНАЯ
ПОЛЕВАЯ
РАБОЧАЯ
СУВОРОВЦЫ
НАРУКАВНЫЕ ЗНАКИ С ЭМБЛЕМАМИ ПО РОДУ ВОЙСК ЧАСТИ

ФОРМА ОДЕЖДЫ И ЗНАКИ РАЗЛИЧИЯ
ПРАПОРЩИКОВ СОВЕТСКОЙ АРМИИ
И ЖЕНЩИН, ПРИНЯТЫХ НА ВОЕННУЮ СЛУЖБУ В СОВЕТСКУЮ АРМИЮ
ПАРАДНАЯ И ПАРАДНО-ВЫХОДНАЯ
ПЕТЛИЦЫ
ПОВСЕДНЕВНАЯ И ПОЛЕВАЯ
НАРУКАВНЫЕ ЗНАКИ С ЭМБЛЕМАМИ ПО РОДУ ВОЙСК ЧАСТИ

ФОРМА ОДЕЖДЫ И ЗНАКИ РАЗЛИЧИЯ ОФИЦЕРОВ СОВЕТСКОЙ АРМИИ
ПАРАДНАЯ И ПАРАДНО-ВЫХОДНАЯ
ПОВСЕДНЕВНАЯ И ПОЛЕВАЯ
КОКАРДЫ И ЭМБЛЕМЫ
ПЕТЛИЦЫ
ЭМБЛЕМЫ НА ПЕТЛИЦЫ И ПОГОНЫ

ФОРМА ОДЕЖДЫ И ЗНАКИ РАЗЛИЧИЯ МАРШАЛОВ И ГЕНЕРАЛОВ СОВЕТСКОЙ АРМИИ
ПАРАДНАЯ И ПАРАДНО-ВЫХОДНАЯ
ПОВСЕДНЕВНАЯ И ПОЛЕВАЯ

Right: Army General V. A. Matrosov, Chief of the Main Department of the KGB Border Guards. This picture was taken shortly before his promotion in 1978, and shows him wearing the three stars of a Colonel-General.

Second right: Head of the leading service of the Soviet Armed Forces, the Strategic Rocket Forces, Chief Marshal of Artillery V. F. Tolubko.

Below: 152mm M-1973 Self-propelled Gun.

Opposite page, top left: Mikhail Sergeyevich Gorbachev, General Secretary of the Central Committee of the CPSU since March 1985.

Opposite, top right: SS-1 Scud-A tactical nuclear missile on the march.

Opposite, bottom: 122mm M-1974 Self-propelled Guns. The Soviet Army puts great stress on the role of artillery on the battlefield.

also directorates concerned with armaments, communications, and coordination of the efforts of the armed forces of the Warsaw Pact. Furthermore, the General Staff examines new developments in technology and their military application. To attain a position on any of the decision-making directorates of the General Staff, an officer must pass the two year course at the General Staff Academy, the highest of all the many military academies in the USSR.

The administration of the largest branch of the SAF, the Ground Forces, as well as that of combat elements of the Air Forces, is carried out by the General Staff through a geographically-organized system of 16 Military Districts within the USSR, and Groups of Forces in East Germany, Poland, Czechoslovakia and Hungary. The other branches – the Strategic Rocket Forces, the Air Defense Forces, and long-range and transport aviation elements of the Air Forces – have their own command structure answerable directly to the Minister of Defense through their respective C-in-Cs, in the same way as the Rear Services Troops and Special Troops. The fifth branch, the Navy, has its own command system. The Deputy Minister of Defense for the Navy controls four fleets and one flotilla, and he, too, reports back directly to the Minister of Defense. The composition of the SAF is complicated still further by the existence of two bodies of troops which do not come under the control of the Ministry of Defense. These are the KGB Troops, who are part of the Committee for State Security, and the Internal Security Troops of the Ministry of Internal Affairs (MVD). Their organization and administration are the responsibility of the respective heads of these departments.

Above: Komsomol members stand guard at the Tomb of the Unknown Soldier in Novgorod. Basic military training is an obligatory part of the curriculum in Soviet schools, and it is considered an honor to be chosen to help defend the monuments of the Motherland.

Above right: Soldiers of the Strategic Rocket Forces rush to their positions in an underground control center. Because of its prestigious position as the leading force, the SRF is amongst the first to choose its soldiers when new conscripts are called up.

The Soviet Services

As has been seen, the Soviet Armed Forces do not split into the three convenient areas of army, air force and navy which we are used to in the West. Furthermore, as well as the usual rivalry found between different branches of the armed forces of any country, there is a definite pecking order of importance established by the Kremlin, and it reflects once again the principal *raison d'être* of the SAF: the need to defend the Motherland.

The Strategic Rocket Forces

The leading service of the SAF, the Strategic Rocket Forces (SRF), is also the youngest. In anticipation of the development of rocket technology, a rocket unit was formed from a Guards regiment of rocket artillery in 1946, under the control of the Main Artillery Direc-

torate. More of these units were formed in the late forties and early fifties, and it would appear that by the mid fifties the Soviet Union had decided to concentrate on the development of the missile as the prime means of nuclear delivery, as opposed to the bomber. Successes in this area were underlined by the first launch of an ICBM in 1957, and the first rocket into space in the same year. This paved the way for the formation of a separate service, the SRF, in January 1960. The tasks allotted to the SRF illustrate why they are so highly regarded and respected. The principal task is the destruction of an enemy's nuclear delivery means, as well as his major military formations and bases. Other targets would include military-industrial complexes, military and civilian headquarters and government buildings, and the enemy's rear services and means of transport. Depending on the circumstances, the missiles of the SRF could be targetted individually, in groups, or in mass concentrations. Because the personnel of the SRF are considered an elite, the selection process is stricter than for the other services, with potential officers having to pass a board of their local military commissariat, as well as the usual examinations. The SRF includes not only the actual missile units and formations, but has its own research institutions, military-scientific

Above: Chemical Troops decontaminating T-64 tanks at a field NBC checking post.

Left: Transporter/launcher of the SSC-1B Sepal coastal defense missile in its firing position. This missile is the land-based version of the shipboard SS-N-3 missile.

establishments, and units of supply and maintenance, too, giving it almost total independence. Conversely, however, it does not have its own arm of service uniform badge, SRF personnel wearing the same crossed cannon as the Artillery Troops.

The Ground Forces

The Ground Forces constitute by far the largest branch of the SAF, and are trained for fighting in both conventional and nuclear conditions. While they lost their position as the principal service after the formation of the SRF in 1960, the realities of fighting a battle on the ground, and the advent of tactical nuclear missiles meant that they would still make up the hub of the SAF. There are four main arms of service in the Ground Forces: Motorized Rifle Troops, Tank Troops, Rocket Troops and Artillery, and Air Defense Troops. Furthermore, because the major role of special troops and rear services is in support of the Ground Forces, these are often included under the same umbrella, although their command structure is different.

Motorized Rifle Troops. With some 130 motorized rifle (MR) divisions, the descendants of the foot-slogging infantryman are still more numerous than any other soldiers in the modern SAF. However, sophisticated technology means not only that the modern Soviet 'infantryman' is transported by tracks or wheels instead of on his own feet, but also that he has combined arms support. Thus, an MR division consists of three MR regiments, one tank regiment, an artillery regiment, an air defense regiment, a surface-to-surface missile battalion, one antitank battalion, and a battalion of multi-barrelled rocket launchers, as well as smaller units and sub-units of special service troops such as engineers and signallers.

Tank Troops. The main striking units of the Ground Forces are the Tank Troops. As well as being used to support the MR Troops, the Tank Troops are organized into 50 tank divisions, with three tank regiments and one MR regiment per division, as well as artillery and service support. It is estimated that there are 50,000 tanks actively deployed in the SAF, from the older T-54/55s to the up-to-date T-80. However, given the Soviets' reluctance to destroy outdated equipment, but rather to put it in mothballs, it is likely that the SAF could field many more tanks than this.

Rocket Troops and Artillery. Missiles of tactical and operational significance (that is, those with a range of less than 1000 kilometers) come under the control of the Ground Forces. The best known of these are the FROG and the Scud (NATO codenames), with ranges up to 70 kilometers and 280 kilometers respectively, and the replacement for the FROG, the SS-21, which has improved accuracy and a range of 120 kilometers. All three can carry either nuclear or conventional warheads. The Artillery Troops have an impressive array of towed guns and self-propelled howitzers at their disposal, as well as antitank guns and missiles, mortars, and the multi-barrelled rocket launchers nicknamed 'Katyusha' when they were first used in the Great Patriotic War. The latest of this family is the 16-barrelled 240mm BM-27.

Ground Forces Air Defense Troops. These troops have the role of defending units deployed on the battlefield and in the immediate rear. Their use reflects the Soviet belief in strength in numbers, as they are capable of a multi-faceted attack on any intruding aircraft, flying anywhere from very low to high altitude. The combination of tube weapons and missiles gives overlapping fire cover, thus making the enemy's task even more difficult. Nevertheless, the SAF boast that today's weapons, especially the surface-to-air missiles (SAMs), are highly accurate, and that one or two SAMs could do the job of bringing down an enemy aircraft that it took about 500 shells to achieve in the Great Patriotic War. Of the tube weapons, the ZSU-23-4 self-propelled anti-aircraft system is highly regarded, with its four 23mm cannon, and the SA-8 Gecko is particularly impressive among the many missile systems.

The Airborne Forces. With their own air-droppable armored personnel carriers (APCs) and artillery pieces, the Airborne Forces are really a service within a service, and are in fact answerable directly to the High Command. Their role is to carry out operations in the enemy rear, particularly the destruction of an enemy's nuclear means, and the capture of key points of communication to assist the follow-up forces. Their effectiveness was shown by the Soviet invasions of Czechoslovakia in 1968 and Afghanistan in 1979, when they paved the way for the main body of forces.

Special Troops and Rear Services. The term 'Special Troops' covers a broad spectrum in the SAF, and refers to troops who have some specific technical skill, such as Engineers, Chemical, or Signal Troops, which they use in support of the main attacking force. The *Engineer Troops* are employed in a combat engineer role, as well as the supporting roles of road-laying, bridging, and obstacle-crossing. The task of the *Chemical Troops* is the reconnaissance of areas affected by nuclear, biological or chemical weapons, and decontamination and disinfecting of the area and also of vehicles and weapons. All communications in the SAF are the domain of the *Signal Troops*, and transport, both in terms of vehicles and the construction of roads and railways is covered by the *Motor Transport Troops*, the *Road Construction Troops* and the *Railway Troops*.

The movement of supplies on these roads and railways is the responsibility of the *Rear Services*. In many cases, the tasks of the Soviet Rear Services are the same as for any army: the provision of food, clothing, ammunition, fuel, spare parts, and medicine being the more important. There are differences, however, in the organization. Firstly, the structure of the Rear Services is split into three definable areas: the strategic rear provides the link with the logistical center, under the command of the Chief of the Rear Services, one of the 10 deputy ministers of defense; the operational rear

Above: Special decontamination of T-64 tanks using heat treatment. This is another task of the Chemical Forces.

Left: Soviet military doctrine emphasises the importance of reconnaissance at all levels. Here combat reconnaissance troops are being assigned a mission. Note the camouflage combat smocks, and the 7.62mm RPK light machine gun mounted on the sidecar.

Overleaf (Main Picture): An early version of the T-54 Main Battle Tank (MBT). Though now superseded by later types of Soviet tanks, these are still in service in Warsaw Pact armies, showing the Soviet Armed Forces' reluctance to dispose of military materiel.

Overleaf, top left: T-62 MBT.

Overleaf, top right: Motor-Rifle (Infantry) Troops deployed from their BMPs, armed with AKM rifles and the AGS-17 grenade launcher.

642

126

Examples of Soviet air power.

Right: Tupolev Tu-95 Bear-D on a strategic reconnaissance flight over the Atlantic.
Below: Although now being replaced in the Soviet Air Defense Forces by more up-to-date fighter-interceptors, the MiG-21 Fishbed has proved immensely popular, not only in the USSR but throughout the world.

carries the supplies to the SAF at Army, Front and Fleet level; and the immediate rear supervises the provisions of stocks to units at divisional level and below. Also, the Rear Services include formations which would not be found in western armed forces such as *Pipeline Troops* and *Construction Troops*, responsible not only for military building, but projects in the civilian economy, too. Though not an element of the Rear Services, the SAF has a final line of support which, again, is without an equivalent in the West: these are the *Civil Defense Troops*. They are controlled by the Ministry of Defense, and act as an interface with, and a professional nucleus of, a large-scale civilian civil defense organization.

Air Defense Forces

Founded in 1948 as a separate branch of the SAF, the Air Defense Forces have vied with the Air Forces for third place in the pecking order, and would now seem to be established in that position. Their purpose is the defense of the population, administrative centers, groupings of forces, and other potential targets, by destroying an enemy's air attack, whilst in flight. The Air Defense Forces are organized into Air Defense Districts throughout the USSR. They have close links with the Ground Forces Air Defense Troops, and also the Civil Defense Troops. There are three distinct areas of the Air Defense Forces: Anti-aircraft (AA) missile troops; Air Defense (PVO) Aviation; and Radio Technical Troops.

For some years now the principal weapons in the arsenal of the AA Missile Troops have been the SA-2 Guideline and SA-5 Gammon. It was an SA-2 which shot down the American U-2 reconnaissance aircraft piloted by Gary Powers in May 1960, which is still held up today as an example of the worth of this arm of service.

Whereas SAMs are essential for the Ground Forces Air Defense Troops, of more importance to the Air Defense Forces is the PVO Aviation, with its fighter and interceptor aircraft, particularly those of the MiG family. The original MiG-21 Fishbed has now been replaced by the Su-15 Flagon, the MiG-25 Foxbat A and E versions, and the MiG-31 Foxhound among others. It was an Su-15 of the PVO Aviation which was responsible for the widely condemned shooting down of Korean Airlines flight 007 on 1 September 1983.

The lynchpins of the Air Defense Forces are the Radio Technical Troops who operate the radars which guide the service's aircraft and missiles. The Soviets claim that their radar operates without interruption round the clock and in all weathers; this is probably not an idle boast, since the Soviet Union possesses tremendous density of coverage, often using older equipment as a back-up for newer technology.

In the early 1980s, the Air Defense Forces went through a major reorganization, which created a unified air defense command, linking the strategic air defense assets with those of the fighting forces, and increasing the operational flexibility of air and air defense deployment.

Below: The MiG-23 Flogger is another fighter-interceptor used by the Air Defense Forces as well as by the Air Forces

Main picture: One of the mainstays of Military Transport Aviation of the Soviet Air Forces is the Antonov An-22 Cock. These are also in service with the civil airline Aeroflot, and could be used to supplement the military's requirement.

Opposite: Strategic bombers still have an important part to play in Soviet military planning. These are the speciality of the design bureau named after A. N. Tupolev, including the Tu-26 Backfire.

The Air Forces

In time of conflict, the Soviet Air Forces would be expected to gain mastery of the air, carry out decisive strikes against groups of enemy forces and their headquarters, destroy military-industrial complexes deep in the enemy rear, give support to their own troops, land and drop troops, equipment and supplies, and carry out air reconnaissance. In order to carry out these missions the Air Forces are organized into three arms: Long Range Aviation; Frontal Aviation; and Military Transport Aviation.

Long Range Aviation is the main striking arm of the Air Forces. It represents part of the Soviet Union's intercontinental nuclear capacity, and, although apparently under threat when Khrushchev was in power, today has over 600 long and medium range bombers, which, with the aid of tanker aircraft, could fly close enough to the continental USA to keep out of range of air defense systems, yet still launch their missiles and return to their bases in the USSR.

Frontal Aviation is designed to support the Ground Forces, by means of bomber, fighter-bomber, reconnaissance, and special aircraft, as well as combat and transport helicopters. The introduction of supersonic fighters armed with tactical nuclear missiles is seen as a major step forward in the fighting capacity of the Air Forces.

Military Transport Aviation includes heavy lift aircraft and helicopters for the transportation of both troops and equipment, such as the An-22 Cock and the Mi-6 Hook. It is this arm of service which provides the transportation for the Airborne Forces. Because of the centralization of control in the Soviet Union, Military Transport Aviation can be supplemented at any time by aircraft from Aeroflot, the civilian airline. This actually happened during the Soviet invasion of Czechoslovakia in 1968.

АЭРОФЛОТ

Above: Soviet servicemen are continually being exhorted to live up to the heroic examples of the past. On this contemporary Soviet poster a sailor flashes up the message 'True to our Military Traditions!'

Right: The Kara Class Cruiser *Nikolaev* fires off a missile. Like all Soviet cruisers she is heavily armed, notably with the SS-N-14 surface-to-surface missile, and SA-N-3 and SA-N-4 surface-to-air missiles.

The Navy

In both pre- and post-revolutionary Russia, the Navy has always been considered the least important service, although given Russia's position as primarily a continental power this is not surprising. It was only in the 1960s and 1970s that the Soviet leadership seemed to become aware of the full role that the Navy could play in the carrying out of Soviet policy, which would seem to be illustrated by the promotion of Admiral Gorshkov to the rank of Admiral of the Fleet in 1968. Although C-in-C of the Navy since 1956, he was in an inferior position to that of his contemporaries in the Army until his promotion. It should be noted, though, that being fifth in the list of services does not mean that the Navy receives the lowest quality conscripts; on the contrary, because the technology they will have to use on board ship is more complex than that encountered by most soldiers in the Ground Forces, the Navy is among the first to select its men.

The Navy is divided into four fleets: the Northern Fleet, based at Murmansk; the Pacific Fleet at Vladivostok; the Black Sea Fleet at Odessa; and the most highy decorated, the Baltic Fleet, based at the old home of the Russian Navy at Kronstadt. As well as these, there is a flotilla on the land-locked Caspian Sea, and a separate base in Leningrad.

The principal arm of the Navy is the submarine fleet, particularly its nuclear-powered and nuclear-armed boats. In all the Soviet Navy possesses 278 submarines of all types, of which 67 carry long-range nuclear missiles, which, with the SRF and the missile-carrying aeroplanes of Long Range Aviation, give the Soviet Union a three-pronged strategic nuclear capability.

Naval Aviation was transferred from the Air Forces shortly before the Great Patriotic War, and now represents a powerful addition to the sea-going vessels of the service, as it includes nearly 400 bombers such as the Tu-16 Badger and the Tu-26 Backfire as well as jet fighters like the Yak-36 Forger and the Su-17 Fitter. Complementing the shore bases, the Soviet Navy now has three operational 'aircraft-carrying ships' (as it calls them) of the *Kiev* Class, which carry helicopters and V/STOL aircraft; the Soviets deny allegations that they have 'aircraft carriers.'

A major role which the Soviet leadership has realized can be carried out by surface ships of its Navy is that of showing the flag, particularly in Third World countries. Given the strict control and discipline exercised over Soviet sailors, which generally produces good behavior ashore, much propaganda value can be made of these visits. As Admiral Gorshkov has said, 'it is impossible to underestimate the ideological significance of these visits'; the Navy is bound to be, therefore, a valuable political weapon in the hands of a leadership whose ideology is convinced of its eventual world-wide triumph.

A small, but elite, force which comes under the command of the Navy is the Naval Infantry. Organized into five brigades, they are equipped with their own APCs, tanks and self-propelled artillery, and are a naval equivalent of the Airborne Forces, given the task of landing as an advance guard to prepare for the main force.

Spetsnaz, KGB, MVD

To complete the picture of the SAF, mention must be made of three bodies which do not fit into the areas covered by the five branches of the Armed Forces.

The first of these, the *Troops of Special Designation (Spetsnaz)* do come under the control of the Ministry of Defense, but are immediately governed by the Main Intelligence Directorate of the General Staff (the GRU). They are undoubtedly the Soviet Union's finest troops, and are given sabotage, reconnaissance, and assassination missions. Although there is rivalry between the military intelligence network (GRU) and the civilian intelligence network (KGB), there are occasions of co-

operation, too, such as the assassination of President Hafizullah Amin of Afghanistan in December 1979. The Spetsnaz wear the uniform of the Airborne Forces, except for those serving in the marine branch (each Fleet has a naval Spetsnaz brigade attached to it), who dress as Naval Infantry.

Troops of the KGB and MVD do not come under the direction of the Ministry of Defense, but, being equipped with APCs, tanks, helicopters and small arms, fully justify the description of armed forces. Uniformed troops of the KGB fall into two categories. Firstly, there are the Border Guards, easily identifiable by their green epaulettes and collar tabs, who not only patrol the borders of the USSR in their APCs, helicopters, coastal patrol vessels, and on horseback, but carry out the mundane task of checking the passport of everyone entering and leaving the Soviet Union. The other role of the KGB Troops is for ceremonial and guard purposes. All important military and political centers are guarded by these troops, most notably Lenin's Mausoleum on Red Square in Moscow, described as 'Guard Post Number One.' These troops have royal blue markings, with the letters GB (State Security) on the shoulders.

Like the Border Guards and the Kremlin Guard, the MVD Troops are often seen by the visitor to the USSR, as their role as internal security keepers covers any major sporting event and any parade or official celebration. They can be identified by their maroon trimmings, and the letters VV (Interior Troops). The casual visitor will, however, never see them in their other role, which is escorting convoys and prisoners to labor camps, where they also act as guards.

Summary

No one can doubt that in a period of less than seventy years since the Great October Socialist Revolution brought into being the world's first state governed by communist ideology, the USSR has created massive and powerful armed forces. Neither is their defensive role in doubt: Russia's history and geography, together with the order of precedence given to the modern Soviet Armed Forces all bear witness to that. However, it would be foolish to ignore the offensive threat that these forces pose, simply because we in the West have yet to see it demonstrated forcibly. Not only have the nations of Eastern Europe and Afghanistan seen this, but the essence of the ideology which drives these armed forces is that the socialist system will triumph over the capitalist one, and that armed force will have a vital role to play in the denouement. Furthermore, the very existence of these massive forces poses a threat in itself, as the thought of what they could do makes them an instrument of coercion. Large defensive armed forces could very quickly be transformed into large attacking armed forces, particularly if one believes that attack is the best form of defense. If the Soviet leadership felt that the time was ripe for continuing the spread of the Revolution westwards, or if they genuinely believed they were about to be attacked, they would command the Soviet Armed Forces to move onto the offensive. Even declarations such as 'no first use of nuclear weapons' would become meaningless; to wait until the other side used them on you, thus losing the vital element of surprise, would be stupid: whatever else may be true of the Kremlin leadership, stupid they are not.

Above: Last-minute checks before putting to sea are the responsibility of the junior officers.

Opposite page, top: The Soviet Union's nuclear-powered submarines are the Navy's principal strike-force, in both the hunter-killer role and the strategic missile role. One of the most deadly of the hunter-killer classes is the Victor Class. Seen here is a Victor III.

Opposite page, middle: The *Moskva* Class helicopter cruisers are equipped with a spacious flightdeck, and carry either the Kamov Ka-25 helicopter, designed for anti-submarine warfare and search and rescue, or the Mi-8.

Opposite page, bottom: Very different from the *Moskva* is this Kresta II Class cruiser. Nevertheless, this, too, carries a Ka-25 helicopter, seen here landing on the tiny landing pad at the stern of the ship.

2. THE WARSAW PACT & EASTERN EUROPE

Introduction

The principal difference between the socialist system as it is in the Soviet Union and as it is in the other six member states of the Warsaw Pact, is that the Russians brought about their own Revolution, with an ideology largely formulated by a Russian (Lenin), which was adapted to suit the Russian situation; whereas Bulgaria, Czechoslovakia, East Germany, Hungary, Poland and Rumania had a foreign ideology thrust upon them by a foreign occupying power, which had not been adapted to suit each country's situation. In this lies the key to the different attitude to Marxism-Leninism to be found in the USSR and in the rest of the Warsaw Pact. Russia has long striven to be recognized as an inventive nation; hence the spurious Russian claims to be the originator of the radio, the airplane and the tank. But in the communist ideology they can genuinely claim to have something which is theirs: they brought about the first socialist revolution, and people worldwide have sought to imitate it. However, this is obviously not the case with the other six European Warsaw Pact members, and so they are bound, at least, to regard Marxism-Leninism in a different way from their Soviet comrades. The positions of Albania and Yugoslavia are different again, and so merit separate examination.

History: Why Eastern Europe looks the way it does

The history of Eastern Europe has for centuries been a turbulent one, and one where political incidents and changes of the map (of which there have been many) have had far-reaching significance. It was events in this part of the world – the assassination of Archduke Ferdinand in Serbia in 1914, and Hitler's invasion of Poland in 1939 – which sparked off both the First and the Second World Wars. By the mid-1930s the eight countries under consideration had evolved to various different positions. Bulgaria, Rumania, Albania and Yugoslavia were all monarchies, as was Hungary nominally, though actually ruled by a regent. Czechoslovakia was a genuine democratic republic, as Poland claimed to be, although its military leadership could hardly be called democratic. Modern East Germany was part of a united Germany then glorying in the title of the Third Reich. This temporary state of affairs began to crumble in 1938 when Hitler annexed most of Czechoslovakia, and its final death knell was sounded when he attacked Poland in the following year. The outbreak of World War II

meant the countries of Eastern Europe having to choose between supporting Germany or opposing her. Hungary and Rumania declared their support in 1940, and both countries supplied troops for Operation Barbarossa, Hitler's invasion of the Soviet Union, in June 1941. Bulgaria, too, came out on the German side in 1941, though, significantly, did not send troops to fight the Russians. Albania and Yugoslavia opposed the Germans and were invaded; the former by Italy in 1939, and the latter by Germany in April 1941.

By 1944, the influence of the Nazis – short-lived in terms of years, much longer lasting in terms of feelings – was being driven out and replaced by communist authority. This came about in two ways: principally by Soviet influence and the actions of the Soviet Army, but also by developments within the Eastern European countries themselves. The post-war imposition by the Soviet Union of the communist system on these countries would have been much harder to achieve without native communists sympathetic to Moscow. Stalin appeared to be doing himself and his cause no favors before the war, by his treatment of foreign communists who came to the Soviet Union to escape persecution. Many were purged more viciously than Soviet citizens. However, after the USSR's involuntary entry into the Second World War, foreign communists were seen as allies in the struggle against Nazism. It was not just leaders who were trained in the ways of Marxism-Leninism in the Soviet Union, but also Polish, Czechoslovak

After the riots in East Germany in 1953, the first major revolt against Soviet control over Eastern Europe came in Hungary in 1956, and it provided the rest of the Warsaw Pact countries with a vivid demonstration of what would happen to them if they tried to resist.

Above: Hungarian insurgents prepare to open fire on Soviet troops.

Above left: Soviet tanks on the streets of Budapest. The Astoria Hotel in the background shows the scars of battle.

Left: Cadets of the Hungarian Military Academy join the insurgents on the streets of Budapest.

In 1968 it was the turn of the Czechs to feel the weight of Soviet military power, following their attempts to build 'communism with a human face.' Unlike the Hungarian uprising in 1956, the reaction of the people of Czechoslovakia in 1968 was one of peaceful bewilderment at what the Soviets were doing.

Above: A Czech youth carries his country's flag across the path of a Soviet T-55 tank.

and Rumanian troops, who fought alongside the Russians in Eastern Europe in 1944 and 1945. It must also be noted that in the immediate post-war period there was a general feeling throughout Europe that it was the old order which had brought about the war, and that therefore a major change of political direction could help secure a safer, peaceful future. The elections in Britain, France and Italy bear witness to this, the last two countries actually having communists serving in their governments.

Thus, there was a certain amount of genuine sympathy for the communists in Eastern Europe; nevertheless, once the local communists had achieved a degree of power, the presence of the Soviet Army could be used to apply the pressure needed to turn the system into a one-party, Soviet-style one, which was the outcome in Bulgaria, Czechoslovakia, Hungary, Poland and Rumania. The procedure was complicated in the eastern part of Germany, because it was only one part of the whole country that was under Soviet control. What was more, its capital, Berlin, deep in the heart of the Soviet sector, was also under four-power control. As the rift with her wartime allies widened, however, the Soviet Union cared less and less for the pretence of cooperation, and the German Democratic Republic (GDR) was established as a separate country in October 1949. Yugoslavia and Albania were the only two countries to set up communist governments without having Soviet troops on their soil. However, Moscow approved of the regime Tito established in Yugoslavia in 1944-45, and, as the one which took over power in Albania in 1944 was heavily indebted to Tito, that, too, could be expected to follow the Kremlin's wishes.

By 1950, not only were communist governments established in Eastern Europe, but an organization had been set up to bring them into closer cooperation; or rather, the Soviet Union had enforced membership of the Council for Mutual Economic Assistance (CMEA) upon them to ensure that they ran their economies the way comrade Stalin wanted. When the CMEA was formed in January 1949, its members were the USSR, Bulgaria, Czechoslovakia, Hungary, Poland and Rumania; Albania joined the following month. The GDR, of course did not exist as a separate country at that time, and thus the notable omission was Yugoslavia. The previous year there had been a major disagreement between the Soviet Communist Party and the Yugoslav Party over the Yugoslav conduct of foreign policy, which ended with Tito being branded a Trotskyite, and started the rift between Yugoslavia and the USSR which has continued ever since. This was the first indication that the countries of Eastern Europe were not always going to follow Moscow's bidding exactly as she might wish. Fears that this might spread prompted purges of the parties throughout the other countries. Although this was carried out by the parties concerned, it met with the Kremlin's approval. However, although this would appear to have established the loyalty of the various parties, there have been a number of demonstrations that have suggested that the people in these countries are less than happy with their Soviet-dominated system of government.

The first example of this came in 1953 in the GDR. Largely in revenge for Soviet suffering in the Great Patriotic War, the Soviet Union made the GDR suffer economically in the late forties and early fifties. In June 1953 the growing unrest of industrial workers over their low wages and long hours spilled over, and there were riots on the streets of Berlin which spread throughout the country, involving over 300,000 people. The Soviet military commander, General Chuikov, declared martial law, and Soviet troops were used to put down the disturbances. Soviet treatment of the GDR eased somewhat after this.

The USSR was directly responsible for the next bout of trouble which shook eastern Europe, too, and this time it was far more serious. In 1955 (the year in which the Warsaw Pact was formed), Khrushchev began the

Displays of contemporary East German military might, during demonstrations to mark the 35th anniversary of the founding of the GDR in October 1984.

Above: East German soldiers on parade.

Left: Soviet-built FROG-7 tactical missiles. Although these are capable of carrying both nuclear and conventional (high explosive) warheads, control over all nuclear means in the Warsaw Pact armies remains with the Soviet High Command.

denunciation of Stalin, which was then expressed at the Twentieth Party Congress of the Communist Party of the Soviet Union in February 1956. The effects of de-Stalinization were felt throughout Eastern Europe, but especially in Poland, Hungary and Albania. In Poland, there was rioting in Poznan in June, because of low pay and shortages of goods in the shops. This was brutally suppressed by Polish troops. In Hungary, the uprising threatened the Party's monopoly of power and Hungary's membership of the Warsaw Pact, and was put down in bloody fashion by the intervention of Soviet troops in November. Albania felt that the Soviet Union had betrayed the true course of socialism, transferred her allegiance to China, and withdrew from the CMEA in 1961. She ceased to play an active part in the Pact in 1962, but remained a member until 1968.

Following the construction of the Berlin Wall in 1961, to prevent the flow of the young and skilful who were leaving the GDR in their thousands (over 3,000,000 East Germans left between 1945 and 1961), there was a period of calm rebuilding in Eastern Europe which lasted until 1968. Then the Czechoslovaks, who felt a genuine Slav kinship with the Russians and whose leaders had forcefully imposed upon the nation the Soviet system of government, began the experiment of building 'communism with a human face.' After lulling the Czechs into a false sense of security – high level meetings between the Soviet and Czechoslovak leadership ended in smiles, Soviet troops which had been taking part in Warsaw Pact maneuvers announced their withdrawal from Czechoslovakia – the Kremlin acted swiftly to crush the movement in which it saw a threat to the whole socialist system. The air traffic controllers at Prague Airport saw nothing unusual in the request by a Soviet Aeroflot aeroplane to land, and granted it without question. However, having touched down, Soviet Airborne Troops began tumbling out, and the aircraft ran on and took off again, going back for more troops. The Airborne Troops seized the airport, then commandeered transport to take them into the center of Prague, where they took control of bridges and centers of communication, thus preparing the way for the tanks which followed soon afterward. It must be borne in mind that, however slick the military operation might have seemed, neither the people nor the Armed Forces of Czechoslovakia offered any resistance, partly because they were so shocked, but also because, as one Czech put it, 'You just can't go against tanks with bare hands.'

Following the invasion of Czechoslovakia, the Soviet Union declared the 'Brezhnev Doctrine,' which basically said that if any other country of the socialist community was to step out of line, she could expect the same fate. This seemed to have the desired effect. Although there were troubles in Poland in 1970, the Poles themselves were able to deal with them, even though it meant a change of Party leader from Gomulka to Gierek. It was another decade before major trouble flared, and this time it proved to be far more serious than anything Eastern Europe had known. Following strikes in the Gdansk shipyards in Poland, there came into being Eastern Europe's first genuine trade union, Solidarity. As the movement grew in strength in 1980 and 1981, so the Party's monopoly of power seemed to be more and more under threat. Trouble in Poland is a great worry for Moscow, because although Poland does not have borders with any NATO countries (unlike Czechoslovakia) the Soviet Union's vital supply lines to its forces in the GDR go through the country, and disruption of these could play havoc with the USSR's preparations for, or carrying out of, a war in Europe. The situation is complicated still further by Moscow's realization that Poles have a long history of fighting to protect what they hold dear. Whilst the Czechs may have been rational enough to realize that you could not go against tanks with bare hands, there was no guarantee that the Poles would think the same way. After all, they had virtually done just that against the Germans in 1939 (they actually mounted old-fashioned cavalry charges, which were just about as ineffective). Thus, intervention by Soviet troops could well have caused a bloodbath. It must have been with a certain amount of relief, therefore, that Moscow greeted the imposition of martial law in Poland in December 1981. Nevertheless, even though martial law was lifted midway through 1983, the Soviet Union realizes that the fire which erupted in 1980 has still not gone out in Poland.

The formation and development of the Warsaw Pact

When the Soviet Union, Bulgaria, Czechoslovakia, the GDR, Hungary, Poland, Rumania and Albania signed the Treaty of Friendship, Cooperation and Mutual Assistance, which established the Warsaw Pact on 14 May 1955, it was claimed that it was in response to 'possible aggression,' particularly from NATO. Although this seems rather feeble, given that NATO had been in existence for over six years already, it is not totally inaccurate, as only nine days earlier West Germany had joined NATO, and something which would genuinely frighten the USSR and her Eastern neighbors would be the thought of a militarily strong Germany. Possibly the Soviet Union would have orchestrated suitable cause to establish a military pact at some time anyway; but, given the significant gains which she has now made since the formation of the Pact, it would have been exceedingly surprising if it had been allowed to lapse in May 1985, as the original terms suggested it might. In April 1985 it was extended for a further twenty years, with an option to continue for ten more.

For the first five years of its existence, it is not too great an exaggeration to say that the Warsaw Pact was almost totally ignored as an organization by its member states. It should be remembered that, the USSR apart, none of its members had significant armed forces at this time, but even so the highest political decision-

Opposite page, top: East German troops carry out a river-crossing exercise.

Bottom: Soviet-designed AT-3 Sagger anti-tank missile, mounted atop a BMP of the East German Army (NVA). There is now no arms industry in the GDR, and they receive virtually all their military equipment from the USSR.

Top left: Of all the Eastern European countries, Bulgaria is the one which follows the Soviet line most closely. This is reflected even in the presentation of the leader of the Bulgarian Revolution of 1944, Georgii Dmitrov. Like Lenin in Moscow, Dmitrov is displayed today in a mausoleum in one of Sofia's main squares, and a permanent guard is kept by specially chosen troops of the Bulgarian Army. The feather on their hats is a reminder of a pre-Revolutionary battle honor.

Main picture: MiG-23 Floggers of the East German Air Defence Forces.

Above: Soviet and East German air and naval forces combine during maneuvers on the Warsaw Pact exercise *Soyuz-81*.

making body, the Political Consultative Committee (PCC) met only three times. Nevertheless, the individual Pact armies were improving in quality of both equipment and manpower, and in 1961 the first major joint Warsaw Pact exercise, *Burya* (Storm), took place. This seemed to mark the beginning of the real operation of the Pact both militarily and politically, which the non-participation of Albania from 1962 did nothing to shake. Throughout the sixties, there was a growing number of joint exercises, and the PCC met far more regularly. The Soviet invasion of Czechoslovakia in 1968, accomplished with the aid of East German, Polish, Hungarian and Bulgarian troops, illustrates the extent to which the other Pact members not only were prepared to act at Moscow's behest, but also were able to militarily. Nevertheless, the Czechoslovak crisis did strain the Pact, and it was partly in response to this that at the meeting of the PCC in Budapest in March 1969 three new bodies were created to help the coordination of the Pact's military structure. These were the Committee of Defense Ministers, which became the supreme military consultative organ, and which meets annually; the Military Council, which analyses the results of exercises and sets the tasks for the coming year; and the Technical Council.

The period 1969-72 saw a large increase in joint Warsaw Pact exercises, with 11 in 1969 alone. Although numbers dropped off again after this, a significant development which occurred in the non-Soviet Warsaw Pact armies was the interaction of the East German, Polish and Czechoslovak Armies without Soviet units. Also since this time the Soviet Union has been supplying its brother-in-arms with more up-to-date military equipment. In 1976 an important new political body was created, the Committee of Foreign Ministers, which meets annually. Since then the major organizational development within the Warsaw Pact has been the restructuring of the Air Defense (AD) forces throughout, as an extension of the USSR's reorganization of her own AD forces in the early 1980s. This has brought the AD forces of each Pact country more tightly under Moscow's control.

As well as the bodies mentioned above, the military organization of the Warsaw Pact includes a supreme command structure which is not only Soviet-run, but reflects the Soviet Union's own military structure. The Joint High Command (JHC) is the top body, charged with 'the direction and coordination of the Joint Armed Forces.' The JHC includes the Joint Armed Forces' Staff and the Military Council, and meets under the chairmanship of the Commander-in-Chief (C-in-C). Currently Marshal of the Soviet Union Kulikov, the incumbent of this post is, and always has been, supplied by the USSR. Similarly, the other key posts of First Deputy C-in-C, and Chief of the Joint Armed Forces' Staff have always gone to Soviet officers. Although each Pact member has an officer on the JHC, there are also permanent 'military representatives': senior Soviet officers assigned to each Warsaw Pact capital responsible for the Soviet military missions attached to units of the other countries. Thus, Soviet control over the Warsaw Pact is almost total.

This is one reason why the existence of the Warsaw Pact is beneficial to the Soviet Union in particular. Another is that it enables the USSR to maintain troops permanently in Eastern

Europe, not only to enhance its own forward security, but also to ensure the loyalty of its allies. It is interesting to note that when Soviet troops became permanently stationed in Czechoslovakia after 1968, although the reason given was that it was for 'the defense of the *borders* of socialism,' the barracks were situated deep inside the country. Furthermore, the USSR has some propaganda value to gain from the existence of the Warsaw Pact, since she can always claim that, the Pact having been set up after NATO, she would disband the Pact if the NATO countries also disbanded. Although she would rather not now do away with the Pact, the advantages of NATO's disbandment would outweigh the Soviet Union's disadvantages of seeing the Warsaw Pact ended.

The Armed Forces of the countries of Eastern Europe

It must be remembered that a major difference between NATO and the Warsaw Pact is the extent to which the senior partner controls the others. Whilst it is unlikely that NATO's Supreme Allied Commander in Europe (SACEUR) would ever be any nationality but American, the USA does not dictate to its allies the military doctrine and tactics they must employ, the equipment they must use, nor the role they would play in the event of hostilities breaking out. However, these are all areas where the Soviet Union does tell the other members of the Warsaw Pact exactly what to do. Furthermore, it is highly likely that if a war was to occur, the Pact would come directly under the command of the Soviet General Staff, thus making it a virtual extension of the Soviet Armed Forces. This is not to say, however, that each Pact army is simply the Soviet Army writ small. There are notable differences between each one in terms of size and quality; they have specific national roles and military traditions; and they are each regarded differently by the Soviet Union. The following order is an attempt to categorize the armed forces of the Warsaw Pact nations in the way that Moscow regards their value to her, and is followed by a look at Eastern Europe's neutrals, Albania and Yugoslavia.

East Germany

Although the Russians have for a long time mistrusted and even hated the East Germans (the attitude that it was *the Germans* who caused the Great Patriotic War still prevails in the USSR), the strategic location of East Germany, and the German tradition for efficiency seem to have persuaded the Soviet leadership that if they have one Pact ally in particular who will not only do as they are commanded, but do it well, then it is the GDR. East Germans realize that the geography of Europe means that if a major conflict was to start between NATO and the Warsaw Pact, they would be in the front line, and thus faced with the choice of kill or be killed. Although doubts must remain as to how East Germans would perform fighting West Germans (and thus, for NATO the opposite is also true), many Germans on both sides now feel that all they have in common with the people on the other side of the fence is their language. Furthermore, 'traditional German efficiency' has come to the fore again in the GDR, and can be seen in the presentation and performance of the armed

Above: River-crossing by Soviet T-55 tanks during a Warsaw Pact exercise.

Above: The honor guard of East German soldiers in East Berlin is not at a mausoleum, but at the Neue Wache, the Memorial to the Victims of Fascism and Militarism.

Top right: Sailors of the East German Navy taking part in the 35th anniversary celebrations in East Berlin in October 1984.

Right: T-55 tanks of the NVA on exercise. As well as these older tanks, the NVA was the first of the Non-Soviet Warsaw Pact (NSWP) armies to be equipped with the more modern Soviet-built T-72 tank.

forces. This is not to say that the GDR has a free hand to run its own military establishment; far from it. Soviet control over the Armed Forces of the GDR is greater than over any other Pact member. Of the 565,000 Soviet troops stationed in the Groups of Forces in Eastern Europe, over two-thirds (380,000) are in the Group of Soviet Forces in Germany (GSFG), and operationally the East Germany Army (NVA) is subordinate to them. The East German Armed Forces are comparatively small, totalling some 170,000, of whom a little over half are conscripts, who serve for eighteen months in the NVA and the Air Force, or three years if they are in sea-going elements of the Navy. Though small, the forces are well equipped with up to date military materiel, including T-72 tanks, Mi-24 Hind helicopters and SS-21 missiles, and they are kept at a high state of combat readiness. They are supplemented by a large number of para-military forces, including 50,000 border guards; state security troops; interior troops; and the workers' militia, which has armored personnel carriers (APCs), mortars, antitank weapons and anti-aircraft guns.

Czechoslovakia

The Soviet invasion of Czechoslovakia in 1968 destroyed a great deal of pro-Soviet feeling in that country, and in the immediate aftermath the extent to which the Czechoslovak Army and Air Force would have carried out Moscow's wishes in any efficient way must have been open to question. However, the presence of five Soviet divisions, comprising 80,000 troops in all, which form the Soviet Central Group of Forces, is a strong reminder to the Czechs that it would be prudent to do what the Kremlin says. Furthermore, like the GDR, Czechoslovakia is in the front line of any possible European

114

Right: NVA infantry on exercise.

Below: Soviet APCs come ashore in Poland during the *Soyuz-81* Warsaw Pact exercises.

Opposite page, top: Czech troops in action during the *Shchit-84* (Shield-84) joint Warsaw Pact exercises held in Czechoslovakia in the summer of 1984.

Opposite page, bottom: A river-crossing during the 1969 Warsaw Pact exercise Oder-Nysa.

conflict, and thus her armed forces realize the need for combat readiness, as failure could lead to their own country's destruction.

The Czech tradition for making military equipment has been encouraged by the Soviet Union (whilst ensuring that it complements Soviet equipment), and it is used not only by the Czechoslovak Armed Forces, but exported to other Pact members, too. Of particular note are the OT series of APCs and the Tatra wheeled vehicles. In recent years the Czechs have also been beneficiaries of the Soviets willingness to update the equipment of some of their Pact allies, the most notable recent acquisition being the MiG-23 Flogger fighter. Like the GDR, Czechoslovakia has been chosen as a site for Soviet SS-20 intermediate-range nuclear missiles, although this was dictated purely by the country's strategic location. As with all nuclear means in the Warsaw Pact's arsenal, the missiles remain totally under Soviet control.

The Czechoslovak Army numbers about 150,000, two-thirds of whom are conscripts serving a two-year term. The 18,000 conscripts in the 60,000 strong Air Force serve for three years.

Poland

For the first 25 years of the Warsaw Pact's existence, the Polish Armed Forces were regarded highly by the Soviet Union. Although it was realized that many Poles are anti-

Russian, there was comfort for the Russians in the thought that the Poles were probably even more anti-German, and so it was considered likely that the thought of West Germans being part of an attacking force against Poland would have tipped their military support in favor of the Pact. Since 1980, however, and the emergence of Solidarity, with all its accompanying trouble for the Soviet system of government, the Kremlin leadership cannot feel sure of the stance which Poland would take in any conflict, save to say that the Poles would fight tooth and nail to protect their homeland. Nevertheless, the USSR must have been encouraged in military terms by the imposition of martial law by the Polish Army in December 1981, and the installation of an Army General, Jaruzelski, as head of government. On that occasion, at least, the Army proved itself loyal to the system.

Of all the non-Soviet Warsaw Pact (NSWP) countries, Poland has always maintained the largest armed forces. They currently stand at some 320,000, a little over half of whom are conscripts. The Army is organized into three military districts, and of the 5 armored divisions, 8 mechanized divisions, 1 airborne division and 1 amphibious assault division, 10 are kept at the highest state of combat readiness. As well as having largely Soviet equipment, the Polish Armed Forces do have items of home-built equipment, such as armored fighting vehicles (AFVs), trainer aircraft, and

minesweepers in the Navy. Conscripts serving in the Army, Air Force, or internal security forces are subject to two years military training; conscript sailors in the Navy are in for three years.

The other troops stationed in Poland are in the Soviet Northern Group of Forces. With a strength of 40,000, they are the smallest of the Soviet Groups of Forces in Eastern Europe, and are comprised principally of two tank divisions. Being behind the immediate front line, though, the strategic importance of Poland is rather for supply lines for Soviet troops and as a forming-up point for reinforcements. These two reasons make Poland crucial in Soviet war plans, if she is going to maintain momentum in any war in Northern Europe.

Bulgaria

Bulgaria is the most pro-Soviet of all the countries of the Warsaw Pact. However, if Moscow finds that encouraging, she must be disappointed that not only is the strategic position of Bulgaria less crucial than that of the northern Pact members, but also, with a total population of only 9,000,000, the amount of effort she can put into military support for the

Above: This cover of the Czech military magazine *Ceskoslovensky Vojak* (Czechoslovak Soldier) shows Czech infantry in action during exercise *Shchit-84*.

Right: Another Czech military magazine *Atom*, a journal of technical equipment, shows a T-62 tank of the Czech Army completing a snorkeling exercise across a river.

Opposite page, top: Czech-designed Aero L-39 aircraft of the East German Air Force.

Bottom: The armed forces of the Warsaw Pact countries help out the civilian economy in a variety of ways, including helping to gather the harvest. Here a Czech soldier becomes a farm laborer.

USSR is limited. Even so, her armed forces of just under 150,000 represent the largest in terms of percentage of population of all the NSWP countries. Military expenditure, however, is low, a fact reflected by the poor standard of equipment in the Bulgarian Armed Forces. They deploy no tank divisions, and maintain in their five tank brigades a large number of Soviet-built T-34 tanks, the type used by the Red Army in the Great Patriotic War. The Bulgarian Armed Forces employ a higher percentage of conscripts than the other NSWP countries, notably in the Navy and the Air Force, where technology generally demands a higher level of competence, and thus more regulars. Some 3000 of the Navy's 8500 personnel are conscripts, as are 18,000 of the Air Force's 34,000. In line with Soviet practice, the periods of conscription are two years for the Army and Air Force, and three years for the Navy.

Hungary
Hungary maintains the smallest armed forces of any of the Warsaw Pact countries; in the whole of Eastern Europe only Albania, with a population about a quarter of the size of that of Hungary, has smaller forces. Furthermore, their level of mechanization is low, with only one tank division, equipped largely with Soviet-made T-54/55 tanks. The principal strength of the 84,000 man Army is in the five motorized rifle divisions, but, like the tank division, even these are not kept at the highest state of combat readiness. Conscription periods are shorter than for most of the Pact countries, too. Soldiers in the Hungarian Army serve for 18 months, whilst those who go into the Air Force

Above: A MiG-21 Fishbed fighter of the Polish Air Forces.

Right: A feature of Warsaw Pact exercises in recent years has been the number that have taken place between the NSWP members. The illustration shows a joint Polish-GDR exercise in North-West Poland in August 1981.

are in for 6 months longer. The Air Force has just 21,000 personnel, and is equipped with Soviet-built MiG interceptors, transport aircraft and helicopters. The 15,000 border guards are more than two-thirds conscripts, and there is also a tiny Danube Flotilla with patrol craft and river minesweepers.

The Hungarian Armed Forces are supplemented by the fourth Soviet Group of Forces abroad, the Southern Group of Forces. This comprises 65,000 troops, and is organized into two tank and two motorized rifle divisions. As with all other Soviet divisions in Eastern Europe, they are maintained at the highest state of combat readiness.

Rumania

Of all the NSWP countries, Rumania is the least friendly towards Moscow. As well as causing some political embarrassment to the USSR on occasions (such as abstaining from voting in favor of the Soviet Union's invasion of Afghanistan at the United Nations in 1980), she has also gone against the Soviet Union's wishes militarily. Rumania was the only Pact member not to send troops into Czechoslovakia in 1968. Not only does she refuse to have Soviet troops stationed on her soil, she also does not allow through access for Soviet troops. Nevertheless, although Rumania's geographical position means that such decisions do not pose a threat to the Pact's security, it would be impossible for the small Rumanian forces to stand in the way of the Soviet Army, should it decide to get from A to B via Rumania. With total armed forces of less than 190,000, Rumania has the smallest

forces per head of population of any country in Eastern Europe. Furthermore, these forces are poorly equipped. Her eight motorized rifle divisions are based on the aged BTR APC, and the majority of tanks in her two tank divisions are T-34s and T-54/55s. With only 16 months of service, conscripts in the Rumanian Army and Air Force have the shortest conscription period in the Warsaw Pact. Sailors serve for two and one half years in the Navy. In line with the rest of the Warsaw Pact countries, Rumanian border guards and internal security troops are armed with AFVs as well as small arms.

Albania

One of the original signatories of the Warsaw Pact, Albania established close relations with China after leaving the Pact in 1968. Then, believing that China after Mao Tse-tung had betrayed Marxism-Leninism, she broke with China in 1978 and has since been receiving very little military aid from anywhere. Because the Albanian leadership (under Enver Hoxha from the time the Communist Party came to power in 1944 until his death in 1985) believes that theirs is the only country still faithful to the tenets of the ideology, and therefore it must be guarded at all costs, there exists in Albania a siege mentality. Although the armed forces number only 40,000, and are the smallest in Eastern Europe, percentage-wise only the USSR and Bulgaria have more people under arms per head of population. A little over half of this total are conscripts, including some women, thus making Albania the only country in Eastern Europe to conscript women. (The

Above: More Warsaw Pact exercises in Poland. A Soviet T-62 tank crosses a river on a self-propelled ferry. River crossing operations are often rehearsed.

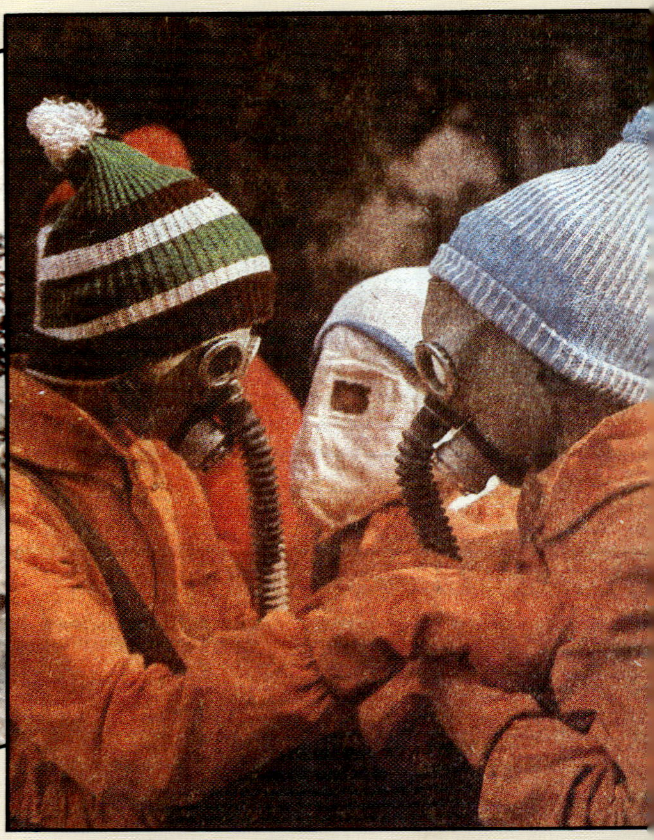

Above: Hungarian artillery in action.

Above right: Civil Defense training is a normal part of school education throughout the countries of the Warsaw Pact. Hungarian schoolchildren are here seen trying out their gas-masks.

Right: A self-propelled gun of the Czech Army crosses a river during the *Shchit-84* exercises.

Opposite page, top: Sukhoi Su-7 Fitter aircraft of the Polish Air Forces.

Bottom: Czech M53 59 Twin 30mm self-propelled anti-aircraft gun.

other countries do have women serving in their armed forces, but they are all volunteers.) Conscription periods are two years for the Army, and three for the Navy and Air Force. The lack of military aid means that the equipment of the Albanian Armed Forces is old and possibly obsolete. T-34 tanks, BRDM-1 scout cars, *Kronshtadt* patrol craft and MiG-15 fighters, all of Soviet manufacture, speak of bygone days when Albania had friends. It is highly unlikely that Albania could mount any offensive action against anyone, even her neighbor Yugoslavia, against whom she still holds territorial claims.

Yugoslavia

Yugoslavia split with the other socialist states of Eastern Europe in 1948, and thus was already pursuing her own non-aligned policy when the Warsaw Pact was formed in 1955. She has never entered into a defense alliance with anyone, although much of her military equipment has been bought from the Soviet Union. This includes tanks, artillery, antitank weapons, naval vessels and systems and some aircraft, although there have been significant contributions by Yugoslavia's own defense industries, too, particularly with aircraft such as those of the Jastreb series. Yugoslavia has the shortest conscription period in Eastern Europe (15 months for all services), and about two-thirds of her total armed forces strength of 240,000 are conscripts. The Army is organized into seven military regions, and its main strength is its ten infantry divisions. There are no tank divisions, only eight independent tank brigades. The Navy is largely a coastal protection force 12,000 strong, and the Air Force includes fighters, helicopters and trainers, as well as controlling the Army personnel of the Air Defense Force. Despite being run by a government which holds to Marxist-Leninist ideology, with all that teaches about the inevitable world-wide victory of communism, Yugoslavia is in no position to help it on its way in the military sense.

Summary

Despite ostensibly being bound together by a common ideology, therefore, it can be seen that there are differences between each of the countries of Eastern Europe. Those further apart from the rest are Albania and Yugoslavia, each pursuing a non-aligned policy, though each in a very different way. It is highly unlikely that either country could ever pose a military threat to NATO, and the size and capabilities of their armed forces suggest that they are totally defensive. The same cannot be said for those countries bound together by the Warsaw Pact, and thus, while individually the NSWP countries pose no greater a threat to NATO than, say, the Belgian Army alone poses to the Warsaw Pact, the Warsaw Pact as a whole, with its anti-western ideology and size of forces, must pose a substantial threat; but to what extent is it united into a whole? A common negative feeling which seems to be true to a greater or lesser extent throughout the NSWP countries (with the possible exception of Bulgaria), is a dislike of the USSR. However, paradoxically, they are brought more firmly under Moscow's control by the resentment and hatred they also feel for each other. The East Germans are universally despised, because the Germans started the last war. Poles and Czechs have disliked each other for many years – far from encouraging the growth of Solidarity in the early eighties, the Czech attitude was that if the Poles did not fall back into line they should receive their just deserts from the Russians, just as happened to the Czechs in 1968. Hungarians and Rumanians still argue over territory on their common border, and protest about the maltreatment of the nationals of the one forced to live in the country of the other. And the Bulgarians are held in contempt for their close relations with the Soviet Union. Thus, the individual NSWP members are not going to unite against the Soviet Union.

How the countries of the Warsaw Pact would react to a command by the USSR to attack NATO is a question which could be answered only if that terrifying event were to happen. However, the peoples of the GDR, Hungary, Czechoslovakia and Poland have seen western inaction in situations where they were being brought very firmly under Soviet control, which would give them little or no hope of western support in the future. And in a situation where the forward-situated countries of the Pact could see that their land would be devastated if they did not try to protect it, self-preservation would be a strong force. It may prove incidental that capitalism and socialism divide the camps; circumstances would seem to dictate that the Warsaw Pact versus NATO would soon become 'them and us.'

Opposite page, left: An Albanian border guard doing his best to ensure that his country remains the most secretive in Eastern Europe.

Opposite page, right: Infantry of the East German NVA on exercise.

Below: Yugoslav pilots prepare to board their MiG-21 Fishbed fighters.

3. CHINA–NEW IDEAS, NEW THREATS

Historical background

The growth of communism in China

At the beginning of the twentieth century China was in turmoil. Corruption and malpractice on the part of officials coupled with an ever increasing western influence inimical to Chinese interests destabilized government and led to an almost universal breakdown of law and order.

At that time few Chinese had received a western education but among those that had there were many with a desperate desire to restore tranquility to the country and build up political stability coupled with economic strength. Many political groups were formed which sought various solutions to the problem ranging from the ideas of liberal democracy through to various socialist philosophies. Marxism, though, was little studied in China until after the turn of the century. The first group of Marxist thinkers established themselves in Peking university only in 1918, the year after Lenin's revolution in Russia. By 1920 many of the group, which included a youthful Mao Tse-tung, had become convinced Marxists. In that year the Communist International (formed in Moscow the year before to coordinate communist activities world wide) sent its first representative to China to establish contact with these Chinese Marxists. Its representative, Voitinsky, worked with the Chinese and as a result the Chinese Communist Party (CCP) was established in 1921.

At first the CCP adopted a very left wing policy, in communist terms, and aimed to work for the immediate establishment of government based on the proletariat, effectively bypassing the capitalist stage of development envisaged by Marx. This radical plan was quickly to be modified, however, to support the course advocated by Lenin for economically backward countries. His view was that communist parties should work with bourgeois (liberal democratic) groups who themselves were fighting to wrest power from autocratic regimes dominated by the landed classes or foreign imperial powers. In this way communists would 'ride on the backs' of liberal democratic governments until they were in a position to seize power themselves.

These ideas were embraced by the CCP in 1922 and by 1924 the CCP had formed a United Front with the Kuomintang (KMT) led by Sun Yat-sen. This group had been the organizers of a revolution which caused the collapse of the Chinese imperial dynasty in 1911. The ideology of the KMT was based on three main tenets: nationalism, democracy and land reform. As such, after 1922, it had received support from

Left: The aftermath of Communist insurrection in 1927. A street in Canton.

Right: A young Mao Tse-tung addressing members of a Kiangsi soviet in 1933.

Lenin, and Russian advisers were active in helping to organize a central KMT government and army with which it could gain control throughout China.

This, the first United Front, lasted from 1924 to 1927 but was never free from dispute between the KMT and the CCP. Sun Yat-sen's death in 1925 led to power falling into the hands of Chiang Kai-shek and a considerable shift of policy to the right within the KMT. Chiang did not sever relationships with Moscow, recognizing the help the Russians might give in his quest militarily to unite China, but he did move against the CCP, toward which he was deeply antagonistic. Many of the senior communist leaders were arrested or assassinated in 1926 and 1927 and yet Stalin (leader of the Soviet Union since Lenin's death in 1924) persisted in supporting the KMT, claiming while doing so that the CCP had deliberately ignored his advice, and effectively caused a rift with the KMT.

In the autumn of 1927 the CCP Central Committee decided that it would make an all out effort to mobilize the urban workers and lead them into the countryside with the aim of drawing the vast mass of the population, the peasants, into line behind them in support of the revolution. Insurrection occurred in several towns but was crushed by KMT troops. In one early battle on 1 August 1927 at Nanchang the CCP forces were quickly defeated but this battle is still celebrated as the occasion of the formation of the Chinese Red Army.

In September 1927 rural uprisings, led principally by Mao, were also organized in the province of Hunan. These, too, failed but nonetheless Mao finally decided in that year that communist victory in China could only be won through the wholehearted participation of the peasantry.

Following the defeat of the Autumn Harvest Uprisings, Mao led the remnants of his forces into the mountains on the borders of Hunan and Kiangsi provinces. There, with some 10,000 men armed only with 4000 rifles he built bases from which gradually to expand his influence. Despite being forced to move from time to time, under threat from KMT troops, he persevered and his strength increased. He was from then onward clear in his own mind that a peasant communist army operating from secure rural bases would provide the key to victory. From 1928 to 1931 he argued his case against stiff opposition within the Central Committee of the CCP where the majority still supported the Comintern, and thus Russian view that the urban proletariat should provide the spark, the impetus for revolution.

By 1931, however, Mao's groups were the only forces achieving any sort of success and at the Congress of the CCP in that year support swung in his favor to the extent that he was even voted Chairman of the party at the meeting. From then onward, despite periodic and often forceful opposition Mao's views were to predominate within the CCP until his death some 45 years later.

The success of Mao's Soviets in Kiangsi and Hunan caused Chiang Kai-shek to move his armies against him and from 1930 to 1934 Chiang mounted one campaign after another into the mountain strongholds. Mao's guerrillas countered these attacks as they were launched with considerable success becoming adept at striking the enemy where and when at his weakest and withdrawing to fight another time when confronted by overwhelming superiority. Mao's theories of guerrilla warfare were largely perfected during these arduous campaigns.

By August 1933 Chiang could muster over a million men against the Red Army and it became obvious that such a weight of numbers would totally swamp the communist held areas. In consequence the CCP Central Committee took the decision to withdraw from the area and move from Kiangsi to a more secure base area. Thus began the 'Long March.'

The main body of the Chinese Red Army, now numbering some 100,000 men, broke through the encircling KMT forces and made its way westward into the remote mountainous regions of the Chinese interior. Thence they headed northward until in October 1935 they reached northern Shensi. They had traveled some 6000 miles through some of the most rugged terrain in the world. Harassed at all times by KMT troops, weakened by the arduous conditions and shortages of food and medical supplies the Red Army was almost decimated by the time it came to a halt. One more assault by the KMT army might well have crushed it completely but other factors came into play at the crucial moment.

Above: Soldiers of Mao's Red Army on the Long March to Shensi in 1935.

A main base was established in northern Shensi: a Soviet was created and the bedraggled remnants of the Red Army were reorganized. The local peasantry were won round to the communist cause and the many recruits which resulted helped quickly to rebuild the communist strength.

At the same time Chiang Kai-shek gathered together a force in Shensi in preparation for a decisive attack against the Shensi Soviet. However, morale in this KMT army was very low. The Japanese were already making preparations to advance into China from Manchuria where they had been in occupation since 1931 and Chiang's generals were loath to commit their army against the Red Army at the moment when the Japanese might attack. The soldiers of the KMT army were, in addition, falling increasingly under the influence of communist propaganda and very reluctant to fight against their fellow Chinese.

Chiang, realizing that all was not well, joined his army in Shensi to ensure that his orders were obeyed. Immediately on arrival he was detained by his own generals and only released following intercession on his behalf by Chou En-lai and other leading communists. They wished to remove the KMT threat from their own army and, seeking to swing all Chinese forces against the Japanese (this conformed with the anti-fascist policy promoted by the Comintern) put forward a proposal for a new united front which was accepted by Chiang and his generals. The Red Army was incorporated into the KMT army, in name at least and the threat to the communists from the KMT was for the time being eliminated.

In July 1937 the Japanese marched into China proper and, sweeping through the coastal regions, reached Canton by October 1938. They made no attempt to advance westward into the interior as their intention was only to lay hold of those parts of China economically beneficial to them. The KMT-led United Front armies, inferior in training and equipment to their Japanese adversaries, fell back before the onslaught until Chiang moved inland to Szechuan province where he established his headquarters at Chungking. Meanwhile the communist forces deployed to protect the Soviet in Shensi. The division of China into these three regions: the Japanese in the coastal areas; the KMT in central southern China and the communists in the interior to the north were to remain largely unchanged until the end of World War II in 1945.

The communists continued throughout World War II to extend their influence further afield throughout the countryside. They won much support from the peasantry because of their very moderate land reform policies and did much to widen their authority by adopting a strong, politically speaking, anti-Japanese stance. The Red Army was ill-equipped to fight the Japanese but it did mount a number of attacks which achieved varying degrees of success militarily, but won over to the communist cause further support from the Chinese people who saw the CCP as the only group trying to defend the Chinese nation from the invader.

In Chungking, the KMT armies went into decline. They were kept inactive because Chiang waited for the Allies to defeat the Japanese hoping meanwhile to retain his armies intact for the ultimate fight with the Red Army which he knew would follow the Japanese collapse. Morale in his forces was further lowered by the powerful propaganda efforts of the communists which often led to large scale defections to their flag. The communists' moderate policies with regard to land reform also strongly influenced the soldiers in Chiang's armies who were, anyway, largely of peasant stock.

The second United Front scarcely existed from the outset in practice and by 1941 not even in name. Waiting for the outcome of the war both communists and KMT prepared themselves in the meantime for the final stages of the civil war.

World War II in the far east came to an abrupt and unexpected end in August 1945 after the dropping of atomic bombs on Hiroshima and Nagasaki. Both factions in China immediately galvanized themselves into action and moved quickly to seize territory, equipment and weapons from the defeated Japanese. The communists in the north moved eastward from their main bases and areas 'liberated' during the war to occupy the rural areas toward the coast. Chiang's forces, with some American help particularly with aircraft, occupied the cities and main road, rail and river communication centers. The Soviet Russian forces which, with Allied agreement, occupied Manchuria in August 1945 withdrew eventually but only to hand over the key Manchurian cities to the KMT. As in China the Chinese Red Army dominated the countryside in Manchuria, however, and they managed to take possession of most of the Japanese arms and ammunition left behind by the Russian troops when they withdrew.

The stage was now set for the final confrontation. From 1946 to 1949 the civil war raged throughout China. The People's Liberation Army (PLA) as the Chinese Red Army was known after 1946, grew in size as more and more of the peasantry swung over to the communist side. The KMT clung desperately to their control of the cities but with their supply lines often cut and the morale of their troops on the decline, one by one Chiang's garrisons

Below: Japanese soldiers execute Chinese civilians during their advance into China in 1937.

surrendered, often with mass defections, to the communists.

In early 1949 Peking was taken by Mao's army and, adopting a more conventional military stance, his guerrillas moved inexorably southward. In April 1949 the PLA crossed the Yangtze river and they were in Shanghai by May. The KMT retreated through Canton and finally collapsed at the end of the year with the remnants of a once enormous army escaping to the island of Taiwan.

The final result of the war was obvious before the end of 1949 and already in October Mao Tse-tung had proclaimed the establishment of the People's Republic of China.

Differences between Mao and the Soviet leadership

In the early twenties, following the formation of the CCP, most Chinese communists saw successful revolution developing along conventional Marxist-Leninist lines with the urban working classes led by the Communist Party taking the lead in the struggle. The Comintern line (the voice of Moscow) confirmed that this was the correct policy but urged the CCP to ally itself with the KMT to advance the cause while at the same time building up the political strength of the proletariat.

In China's desperately undeveloped economic state in the twenties the urban working population formed only a minute fraction of the total which was made up predominantly of peasant farmers. It was this lack of a strong proletariat that led Mao Tse-tung's thinking to diverge from most members of his party and certainly from the Comintern line. His experiences between 1925 and 1927 caused him to work flat out for a revolution through the peasantry. Therefore, for him, the aim was constant – to build an army of peasants to create a communist state. Such a stance was clearly in opposition to the ideas of Marxism and to the experience of the Russians themselves.

Above: A Kuomintang army patrol in action during World War II.

Mao's later experiences in building Soviets in Hunan-Kiangsi in 1930-33 and after the long march to Shensi in 1936 led him to evolve other theories which ran counter to the Soviet Russian view. As regions fell under the influence of Mao's forces he established People's Soviets (Councils) to govern them but his idea was not rigidly to control these Soviets from the center but rather to allow them considerable local autonomy: each Soviet should be independent not only politically but also economically, responsible for allocating its own resources. The Soviet Union's experience had been very different and though in theory the authority of local Soviets in the Soviet Union is considerable, over the years a rigid bureaucracy has evolved tending to exercise ever more centralized control.

Another critically important difference between Mao Tse-tung and the leadership of the Soviet Union stemmed also from Mao's experiences in the thirties in building communist power in the remote regions of Hunan-Kiangsi and later Shensi. Under threat at all times from KMT forces, Mao developed his theories of guerrilla warfare to defend these fledgling Soviets. They met with considerable success. His tactics suited the rugged territory in which his army operated: mountainous regions with scarcely any roads or tracks and thus little opportunity for vehicular movement or the deployment of heavy weapons. Small groups of men, armed with rifles and some machine guns, striking the enemy when least expected, achieved the best possible results in such conditions against regular units of the KMT forces. Further, Mao's army was limited in size. He only had the men from within the area of each Soviet on which to draw and many of them were needed for urgent tasks in other spheres of the normal life of the Soviets. Thus developed the theories of the people's war, fought by an army of volunteers who took up their weapons only when necessary and otherwise carried out a normal civil function.

These ideas of the people's war and volunteer armies ran quite counter to the Soviet Union's experience. From 1917 onward the Soviet Army fashioned itself on the European continental armies. The necessary manpower was found through universal conscription for at least two years and to the best degree possible the army was organized, equipped and trained to fight conventional land warfare.

Below: Soviet troops occupy Harbin in Manchuria at the end of World War II.

Left: US officials meet Mao's leadership in Yenan in 1946 in an attempt to halt the civil war. From left, Chu Teh, Mao, Yeaton, Hurley, Chang Chi-chung and Chou En-lai.

Both the Russian and Chinese communists created their armies in the light of their own experiences. From 1917 onward Lenin, and later Stalin, saw the Soviet Union surrounded by enemies capable of waging war with large conventional armies and it was seen as necessary for the Soviet Union to build a similar army with which to defend itself. The Chinese on the other hand were forced to prepare themselves to fight a civil war against an internal enemy who controlled the cities. A guerrilla army was best suited not only to defend remote firm bases as in the thirties but also to operate in the countryside in the later stages of the civil war to cut communications and ambush the enemy as he sought desperately to maintain his position in the towns.

Mao had disagreed with Stalin after 1924 on many occasions but, though of considerable moment at the time, these disagreements were capable of being ignored for the greater good of communism internationally. Thus, no sooner had he been victorious in 1949 than Mao hastened off to Moscow to sign a treaty of friendship and alliance with Stalin in February 1950. He did not revere the name of Stalin in the same way as perhaps he did Marx and Lenin but he certainly accepted him as the natural leader of the world communist movement. In signing the treaty in 1950 he effectively acknowledged Stalin's paramount position and both of them no doubt saw the new found unity of the two countries as a giant step forward. However, the end was near for Stalin and his death in 1953 was to usher in a new era in Sino-Soviet relations.

The Sino-Soviet split

After Stalin's death there followed a period of inter-regnum in the Soviet Union during which various of his erstwhile lieutenants jockeyed for power. Khrushchev immediately took over the post of General Secretary of the Communist Party and it came as little surprise when in 1956 he started to establish his position as the undisputed leader of the Soviet Union: he was also to become Chairman of the Council of Ministers of the Supreme Soviet (Prime Minister) in 1958.

Initially the period of inter-regnum saw little change in the relationship between the USSR and the PRC but at the 20th Congress of the Communist Party of the Soviet Union in Moscow in 1956 a rift began to emerge. It stemmed from the speeches made by Khrushchev at the congress, none of the texts of which had been distributed to other communist parties beforehand. In one he completely destroyed the myth of Stalin, blaming him for all the problems faced by the Soviet Union since Lenin's death even including the horror and destruction experienced by the people of the Soviet Union in World War II.

This was serious enough but in a further speech he offered new interpretations of Marxism-Leninism which effectively revised many generally accepted elements of the theory. In brief, he suggested that war between the Soviet Union and the West was no longer necessarily inevitable; that advanced capitalist societies might well evolve toward communism without bloodshed and that rather than supporting only communist groups in the Third World the communist powers should offer help to any group seeking to throw off the imperialist grasp.

The Chinese took grave exception to all of this. Mao insisted that the West would not fall without a struggle. He also pressed Khrushchev to adopt an even more aggressive stance toward the West in view of the recent rapid advances in nuclear expertise in the Soviet Union.

Main picture: Chinese Kuomintang soldiers await the arrival of General Marshall in 1946. His was a vain attempt to halt the civil war.

Above, inset: Chou En-lai, Mao Tse-tung and Chu Teh waiting to greet General Marshall.

In 1958 the Soviet Union failed totally to support the Chinese offensive aimed at removing Chiang Kai-Shek from Taiwan and, worse, in 1959 Khrushchev visited President Eisenhower in the US thereby seeming to affirm his determination to live in peaceful coexistence with the West. In the same year Khrushchev visited Peking in an attempt to persuade Mao to moderate his attitude toward Taiwan and to tone down his threatening posture toward the US. Last, but not least, in the same year the Soviet Union reneged on its promise to supply the Chinese with information on nuclear technology.

Now the rift began perceptibly to widen. Mao launched a bitter and vituperative propaganda campaign against the Russians in 1960 and at international communist meetings, indeed, wherever possible, the Chinese attempted to influence others to adopt their line and go against the Soviet 'revisionists' and 'socialist imperialists' as the Chinese now began to describe the Soviet Union.

In 1962 the PRC criticized the Soviet Union's handling of the Cuban missile crisis notwithstanding that it seemed to be an attempt by Khrushchev to get tougher with the West. In the same year the Soviet Union kept out of the

Sino-Indian dispute and immediately after it even began supplying India with military aircraft and other equipment. From 1962 the far eastern Sino-Soviet frontier saw clashes between Russians and Chinese border guards culminating in 1969 in fighting, at least at regimental level, over disputed islands in the Ussuri river.

There have, since the late fifties, been several attempts to heal the breach with little or no success and today the rift remains as wide as ever. The seventies have further seen a shift on the part of the PRC to closer links with the US, Western Europe and Japan which have tended in a variety of ways to exacerbate her relationship with the Soviet Union.

The breakdown in relations between the USSR and the PRC is of great importance to the western world and yet it is extremely difficult to come to any really firm understanding of why it has come about. Is it essentially an ideological disagreement over the interpretation of the basic elements of the Marxist-Leninist philosophy? It is hard to believe this as neither the USSR nor the PRC appear faithfully to have adhered to the tenets of the doctrine but rather to have modified them to suit their own purposes.

Did the whole clash develop out of Mao's undoubted irritation when, after Stalin's death, the new Soviet leadership appeared totally to ignore his claim to be consulted over the future of the international communist movement? Mao was, after all, the undisputed leader of the only other major communist power in the world. It is on record that he was highly critical of Khrushchev not only for failing to bring him into discussions on the future of the movement in 1956 but also because he saw Khrushchev as having no intellectual claim to follow in the steps of the great philosophers and organizers. Undoubtedly, Mao considered himself to be possessed of these qualities and was, it is clear, totally affronted.

It is also necessary to stress, of course, that the period of Sino-Soviet friendship that seemed to spring from the 1950 treaty and scarcely outlived Stalin's death in 1953 was the only time in the history of the two countries when they were not fearful and suspicious of each other's intentions.

The Russians have always been deeply conscious of their European heritage and have further seen their role as a bastion of defense against the Asiatic hordes. The Chinese, on the other hand, have a different view of history. For them all misfortunes that have befallen China have been at the hands of the Europeans culminating in the disastrous experiences of the nineteenth century. Proud of their historic, civilized past they feel it was destroyed by the selfish intervention of the Europeans. Deeply conscious of the continuity of history the PRC cannot forgive the USSR for continuing to cling to territories in East Asia seized originally by Russian tsars from Chinese emperors.

Opposite, below: Communist troops on the march during their final victorious campaign in 1949.

Opposite, top: The successful Communist Revolution in 1949 brings brutal retribution to a Chinese landlord.

Above: Two Communist soldiers surrender to a British soldier of the UN Forces in Korea.

Left: A group of Chinese soldiers taken prisoner by the US Marines in Korea in 1951.

Right: Type 69 main battle tanks shown on a ceremonial parade in 1982. This type of tank was developed from the Soviet T-54/55.

Bottom, far right: Soldiers of the Chinese People's Liberation Army on the Vietnamese frontier in 1983.

Below right: PLA soldiers of a tank regiment taking the oath of allegiance to party and government in Peking in 1981.

Whatever the explanation for the rift it persists and certainly significantly affects the PRC's contemporary view of the world.

Chinese policies and strategies today

The Sino-Soviet dispute and the rapprochement in the seventies between the PRC and the US and the western world has led to a triangular situation developing in the relationships between the PRC, the USSR and the US.

Toward the end of his life Mao had sought to take the PRC along its own path, castigating both the USSR and the US as 'hegemonists' and as 'socialist' and 'capitalist imperialists.' He strove to build up the PRC's capacity to defend itself against what he perceived as a threat from both the other powers. The PRC also saw and continues to see itself to a lesser extent, as the obvious and natural leader of the under-developed world. Her own economy remains backward, sharing the same problems and yet she is influential in international terms as a major communist power.

China has good reason, or so she feels, to fear the threat posed by America and the Soviet Union, particularly the latter as the Russians have developed close ties with Vietnam. However, the PRC found it impossible to build up the necessary industrial and thus military capacity to feel secure from these threats without seeking foreign assistance. As a result the last 10 to 15 years have witnessed a gradual shift away from what has been effectively an isolationist policy to one of closer ties with the western world. The PRC might have sought help from the USSR but has chosen what it feels to be the lesser of the two evils in an attempt to resolve its dilemma. The moves toward rapprochement have always been cautious and very limited but have nonetheless done much to reduce tensions between China and the West. They have also brought considerable benefits to China in terms of the acquisition of the modern technology so vital to broaden and develop her economic base. Recognition by the West has also allowed the PRC to exert more influence within the international community: she now has embassies and missions in all the major capitals of the world; is a member of the United Nations and also has a coveted seat on the Security Council of the UN.

The Forces

The army, navy, air and nuclear forces of the PRC are collectively known as the People's Liberation Army (PLA) and together number some four million men and women. That the whole should be called an army reflects the origins of the force and the predominance of the ground forces which account for approximately 3,600,000 of the total of 4,450,000.

The Chairman of the CCP is ex officio the Commander-in-Chief of the PLA. He carries out his functions of command by chairing the Central Military Commission (CMC) of the Central Committee of the CCP while the forces are controlled on a daily basis through three distinct channels: the General Political Department (GPD); the General Staff Department (GSD) and the General Rear Services Department (GRSD).

The territory of the PRC is divided into 11 Military Regions which provide administrative support for military formations in peace time but which would assume command functions on the outbreak of hostilities.

The strategic doctrine adhered to by the PLA is almost totally geared to Mao Tse-tung's 'military thoughts' which were developed over many years of practical experience of war. They incorporate concepts first enunciated by classical Chinese strategists as well as the ideas of more recent proponents of the art including Clausewitz and Napoleon. In essence the doctrine is based on a defensive posture and is generally referred to as the 'people's war.'

During the prosecution of his guerrilla campaigns in the thirties Mao identified what was for him a factor of critical importance – that the holding of ground was an irrelevance. Today, this notion is still adhered to whenever possible. Should an external enemy attack, the PLA would wish to withdraw, allowing him to advance until he had over-extended his lines of communication and tied down large numbers of troops in occupation duties. Then, when at his weakest the PLA would counter-attack, catch him off balance and drive him from Chinese soil.

In more recent times, however, modifications have been made to this fluid concept of warfare. The experiences of the later years of the civil war, in Korea and in the wars against India and Vietnam have all demonstrated a need to be effective in the more conventional

phases of positional warfare making full use of heavy weapons and armor. Further, the gradual build up of important industrial areas and major cities in China has called the people's war concept into question. They are too valuable to be abandoned and must be defended against an invading enemy. The idea of static defense has thus had to be assimilated into Mao's theories. Thirdly, the advent of the age of nuclear war, and particularly the use of tactical (battlefield) nuclear weapons has required considerable revision of what was after all a rather over-simplified view of the art of war.

Nonetheless, it remains the case that war is the responsibility of all the people of the PRC and not just that of a professional military elite. The organization and functioning of the ground forces of the PLA amply confirms that this is so.

The Ground Forces

The Ground Forces are divided into three distinct components: the Main Forces, the Regional Forces and the Militia. The Main Forces are the best organized, trained and equipped and their primary task is to engage in battle any enemy attack directed against the PRC. Until recent years the Main Forces have been distributed generally throughout the Military Regions of the PRC, positioned to counter-attack any deep thrust by an enemy. Today, a larger concentration of Main Forces is to be found in the Peking and Shenyang Military Regions indicating an intention to hold those areas through a forward defensive strategy.

The Regional Forces are organized to defend their own immediate locality, to provide short term support to Main Force formations in the same area and to organize and lead guerrilla units formed in wartime to harrass the enemy's rear.

The third, element, the Militia, provides the only reserve available to the Ground Forces. In peacetime the reserve lists are maintained so that all the people can be enrolled on the outbreak of hostilities in defense of their own home localities.

The Main Forces are organized into some 42 armies. Most of these armies are of what is known as Type A which are the best equipped. The remainder are of Type B and they have far less armor and fewer vehicles and instead are equipped with horses and mules and pack artillery. This allocation of weapons and armor indicates something of a shortage: the extra equipment needed is just not available. At the same time the lighter weight Type B armies are almost invariably located in more remote parts of the PRC where animals are useful and wheels are not.

The armies of both types are basically infantry formations. Each comprises three infantry divisions in each of which there is one tank regiment. To provide additional armored weight there are twelve independent armored divisions which are distributed throughout the PRC to provide added punch to the armies in what are considered particularly vulnerable areas.

Below: A Luta Class guided missile destroyer of the PLA Navy refueling at sea during exercises in 1980.

Command of the Main Forces is direct from Peking in peacetime but in war they would be controlled by the Military Region Headquarters and be formed, together with Regional Forces and mobilized Militia, into a Front whose whole efforts would be directed from the Commander-in-Chief's headquarters in Peking.

The Regional Forces are many fewer in number than the Main Forces and for the most part comprise independent divisions, regiments and battalions. All these formations and units are far more lightly equipped than the Main Forces. As already stated their wartime role is to support the Main Forces, to engage in local defense and to organize and train the Militia and in particular, to build up guerrilla units. In peacetime they engage in a certain amount of training but much of their time is taken up with recruiting, recruit training, administering and training the Militia and frequently providing man power to assist in industrial and agricultural enterprises within their Region.

Today the Ground Forces of the PLA are on the whole well trained and dedicated to the defense of China and the CCP. Despite some severe shortcomings in weapons and equipment, they are better equipped and led than ever before and since the traumas of the Cultural Revolution have developed into a considerable force.

While China strives to build up its industrial base it remains the case that there are grave shortages of certain military equipments within the Ground Forces. The production of high quality artillery pieces has been stepped up in the seventies and eighties and the PLA now possesses three times as many guns as the United States for example, although this still means many fewer per division bearing in mind the huge size of the forces.

The PRC currently produces a number of types of tanks but all are considered to be obsolete and in view of the fact that the Soviet Union has deployed T-72 tanks in the Far East the PLA must be considered at a serious disadvantage in this particular field. The Chinese can muster some 11,000 tanks in all (a fifth of the total Soviet tank force, though most of the Soviet Union's armor is deployed in Europe).

Transport remains a problem for the PLA. While Type A armies possess numbers of armored personnel carriers it is the case that most of the infantry will march into battle on foot or, for lengthy moves, rely on the rail system.

For antitank defense the PLA continues to use mines and conventional antitank artillery and at low level, in infantry units, the RPG-7. In the late seventies the PRC started bargaining in the West to buy a more sophisticated antitank missile. A vehicle mounted equipment has been purchased but as yet a light weight infantry weapon to replace the RPG-7 has not been obtained.

Efforts are similarly in hand to acquire surface-to-air missiles to boost the effectiveness of the ground forces in protecting themselves from enemy air attack. At present anti-aircraft machine guns and light anti-aircraft guns can provide limited protection against low level attack but little else.

The PLA is equipped with a small number of surface-to-surface missiles with conventional warheads but as far as is known it has no tactical nuclear missile within its inventory of weapons.

Above: Fast gun boats of the PLA Navy on exercise in the Yellow Sea.

The Naval Forces

The PLA Navy (PLAN) comprises some 1500 naval vessels and fighting ships with an air arm of approximately 500 aircraft. The total manpower of the force including the naval air arm and coastal defense forces is believed to be roughly 360,000. The 1500 ships are made up of some 13 destroyers, 21 frigates and 101 submarines. The remainder includes a wide variety of small craft: motor gun boats, hydrofoil motor torpedo boats, conventional motor torpedo boats and missile armed patrol boats. The naval air arm possesses a few intermediate and medium range bombers but consists primarily of fighter aircraft.

PLAN is controlled ultimately by the Commander-in-Chief of the PLA and its day to day functioning is directed by the three staff departments of the PLA. The influence of the navy is much less than that of the ground forces largely because it is so much smaller but also because it has few representatives within the higher echelons of the CCP and certainly little clout in the Central Committee.

Direct control of the navy is exercised by a naval headquarters in Peking and the force is divided into three fleets. The North Sea Fleet is based at the port of Tsingtao and operates off Peking and in the Yellow Sea. The East Sea Fleet is based at Shanghai. It protects Shanghai itself and patrols in the Taiwan Staits area. The South Sea Fleet has its headquarters at Chan Chiang and patrols the PRC's coastline south to Vietnam, it also affords protection to the active commercial port of Canton. The headquarters of the three fleets have operational command not only of all their fighting ships but also of all naval air units, all shore based facilities and coastal defenses, too.

Above: Sailors of the PLA Navy receiving instruction from their ship's political officer at the time of the Cultural Revolution. Each sailor enthusiastically waves his copy of *The Thoughts of Chairman Mao*.

The role of the PLAN is primarily the defense of the PRC. It is responsible for the protection of the Chinese coastline against major naval or amphibious assault and is also organized to patrol the coastline closely to apprehend attempts made by small groups to infiltrate the country.

PLAN does not possess the necessary specialist ships to mount major amphibious operations though small landings could be put into effect with the few landing ships and craft that are available. Some comparatively local protection can be afforded to commercial shipping in the approaches to Chinese waters but the navy lacks the capability to operate in the oceans of the world: the major fighting ships are mostly old and there are insufficient vessels to maintain a fleet on the high seas for any appreciable length of time. PLAN could operate in its coastal defense role reasonably effectively but here again its flexibility is hampered by the need to rely for air cover on shore based aircraft.

The original fleet in 1949 was made up of British and US vessels captured from the Nationalist forces, all of World War II vintage. These were augmented in the fifties by acquiring a number of Soviet fighting ships and others which were built on Soviet designs in China. Few surface fighting ships have been built since the early seventies and thus most are now obsolescent. Guided missiles are fitted to the newer ships and have been added to a few older generation vessels.

The majority of the comparatively large fleet of submarines are also obsolete but the building programme continues and less noisy and more effective underwater vessels join the fleet each year. So far as is known there is only one ballistic missile firing submarine in service and this is only at the experimental stage. There are believed to be only two nuclear powered submarines at present in the inventory.

Above: Military aircraft under construction in factories in the PRC. Chinese copies of the MiG-21 and Tu-16 can be seen.

The large number of small fast patrol vessels of varying types makes up a major element of the fleet and new craft come into service each year to replace older models which then go out of service.

The major weaknesses of the PLAN comprise surface-to-surface missiles which are out of date and capable of being blinded by sophisticated equipment readily available to potential enemies; no effective missile to protect ships from air attack; little, if any equipment to detect or deal effectively with enemy submarines. All of this limits the fleet to operating within reach of cover from shore based aircraft.

Despite these limiting factors it would be possible for PLAN to achieve its aim to defend the Chinese coastline, if only for a limited period, through sheer weight of numbers. Further, if current priorities are maintained, each year will see a considerable improvement in the naval forces' general capabilities.

The Air Force

The PLA Air Force (PLAAF) is very large: in size it is second only to those of the USSR and the US. It comprises something in the order of 6100 aircraft including approximately 4000 fighters, 500 ground attack fighters, 120 medium bombers, 580 light bombers, 550 transport aircraft and approximately 350 helicopters. The PLAAF controls all air defenses, including surface-to-air missiles, anti-aircraft artillery, early warning systems and also the airborne forces. The total manpower of all these elements is estimated to be in the region of 490,000 personnel. As with the ground forces and the navy, ultimate control of the PLAAF is exercised by the Commander-in-Chief of the PLA through the Central Military Commission. The GPD, GSD and GRSD control and provide administrative support down through PLAAF headquarters.

Below the headquarters the air forces are organized on the basis of 10 air districts which correspond to the ground forces Military Regions (there is no air district in the Shenyang Military Region). Within each air district the air forces are subdivided into air divisions which are triangular in breakdown with 3 regiments of 3 wings of 3 squadrons in each. Certain aircraft of a specialized nature or of considerable obsolescence are organized into independent regiments answerable direct to air districts.

Surface-to-air missile and anti-aircraft artillery regiments are allocated to air districts according to need and all early warning radar systems are coordinated by air defense sectors which come under the aegis of air defense zones which answer to the air districts.

The role of the PLAAF is primarily the defense of the PRC and to that end the bulk of the force consists of interceptor fighters which

Above: PLA airborne troops train in the use of helicopters during a large scale exercise in 1981.

are deployed throughout the air districts to respond to intruder aircraft spotted by the early warning systems. There is limited support available for the ground forces in the form of ground attack fighters and an air reconnaissance and transport facility is also available for ground forces use. The bomber force has a nuclear capability with both medium and light bombers.

Despite its considerable size PLAAF suffers from serious technical limitations in virtually all areas. It would be hard pressed to respond to, or even survive, an attack by sophisticated forces such as those of the Soviet Union or the US though it could certainly hold its own against other air forces within Asia.

The PLAAF is equipped almost completely with obsolescent aircraft. The technical expertise is almost certainly available to design and produce modern aircraft but the costs of development are prohibitive in terms of the struggling economy of the PRC. As a result the best that has been done is to modify existing aircraft designs to update them and this has led to the production of a few comparatively sophisticated machines albeit far inferior in capability to the latest models of the Soviet Union or USA.

In the heyday of Sino-Soviet cooperation in the early fifties the Soviet Union supplied aircraft to the PRC and also provided experts to establish a domestic aircraft manufacturing industry. From this original connection the inventory of PLAAF aircraft has derived largely from Soviet designs. The bulk of the interceptor fighters and ground attack fighters comprise MiG-15, -17, -19 and -21 aircraft but since the late seventies new designs based on the MiG-23 (an aircraft of this type was obtained from Egypt) have been developed.

The main defect of virtually all of these aircraft is that they lack a sophisticated all weather capability. Generally speaking navigation aids in most types of aircraft are very primitive and in consequence in poor weather conditions they do not fly.

Although the transport fleet may seem large it is nowhere near sufficient to allow for major troop or supply movements should they be necessary. Indeed, it is doubtful whether there are sufficient aircraft available to fly the four divisions of airborne troops in one lift and they certainly would not be able to move the necessary heavy equipment for combat.

The air defense system is also obsolete relying as it does on the Soviet SAM-2 system. Modern US and Soviet aircraft are equipped to avoid or neutralize such weapons.

In summary the large force available to the PLAAF could make an attack against it a costly affair even for the USSR or the US. The force would, however, be quickly neutralized. The PLA leadership is only too well aware of this and is actively seeking to buy technology from the West to rectify the situation. The shopping list includes vertical/short take off and landing ground attack aircraft, modern jet engines and air launched antitank weapons.

Nuclear Forces

The origin of Chinese nuclear weapons can be traced back to the late fifties when, as a result of a defense agreement signed with the Soviet Union, Soviet nuclear expertise was offered to the PRC. Following the Sino-Soviet rift from 1960 all Soviet assistance, including technical experts, was withdrawn but by then the Chinese possessed the 'know how' to continue to develop their own nuclear program. The first atomic device was exploded in 1964 and this was followed in 1967 by the detonation of a hydrogen bomb. Since then steady progress has been made.

While it does not compare in size and sophistication with western nuclear arsenals the Chinese nuclear force has developed until today it claims to be an effective deterrent against any potential aggressor. The claim certainly holds good when related to other Asian powers but is open to challenge in relation to either the Soviet Union or the US.

It is generally accepted that the PRC possesses some 50 Medium Range Ballistic Missiles (MRBM); 10 Intermediate Range Ballistic Missiles (IRBM) and perhaps 4 Intercontinental Ballistic Missiles (ICBM) with a range of 13,000 kilometers. The PRC also has an unknown number of nuclear bombs which might be delivered by its fleet of obsolete bomber aircraft. On current estimates it is believed that the PLA does not possess any tactical nuclear weapons though if limited resources could be diverted to this field the expertise is available to go into production.

All nuclear weapons are controlled tightly from Peking and are organized within a formation known as the Second Artillery (the PRC equivalent of the Soviet Strategic Rocket Forces) which comes directly under the authority of the Central Military Commission. Little if anything is known of the organization of the Second Artillery at lower levels.

The role of the PRC nuclear forces is to establish a deterrent against would-be aggressors. China has repeatedly pledged that she would never be the first to use nuclear weapons in a war and this statement raises serious doubts as to the effectiveness of the weapons as a deterrent against either the Soviets or the US.

The limited number of weapons remains a serious weakness when the enormous nuclear arsenals of the USSR and the US are considered. The Chinese nuclear target acquisition capability is also open to question and the accuracy of the missile guidance systems may be a major weakness. Indeed, the leadership of the Soviet Union may well consider that all these negative factors add up to nullifying the deterrent. On the other hand, year by year the Chinese deterrent improves and it will become more and more effective.

Below: A PLA Air Force unit under instruction. The photo taken in 1964 depicts already obsolete bomber aircraft.

Combat experience

Since 1949 the PLA has fought in a number of military campaigns from which it has gained considerable experience and modified its overall strategy, tactics, training and administrative procedures. Today's senior leadership will almost certainly have seen service in Korea, as will many current regimental and battalion commanders in Vietnam, and the PLA will have gained greatly from such experience.

Korea
The Korean War began in the summer of 1950 with an attack by North Korean forces against South Korea. Following the intervention of United Nations forces when it appeared that the PRC itself might be threatened as the UN force drove northward back into North Korea, the PLA was forced to intervene.

Only a year after the Civil War in China had ended the PLA was badly equipped and trained to engage in conventional war against a well-armed, sophisticated army. Initially, some 300,000 soldiers of the PLA entered Korea to join the war. They were poorly prepared, not even every man had a personal weapon and there was scarcely any transport and no artillery or tanks. Despite this, the use of the massed human wave attack initially caused the UN army to disintegrate. It fell back before the Chinese hordes who, moving on foot and by night, could never be detected until they fell upon units with complete surprise. The Chinese People's Volunteers, as the PLA formations in Korea were known, used the tried guerrilla

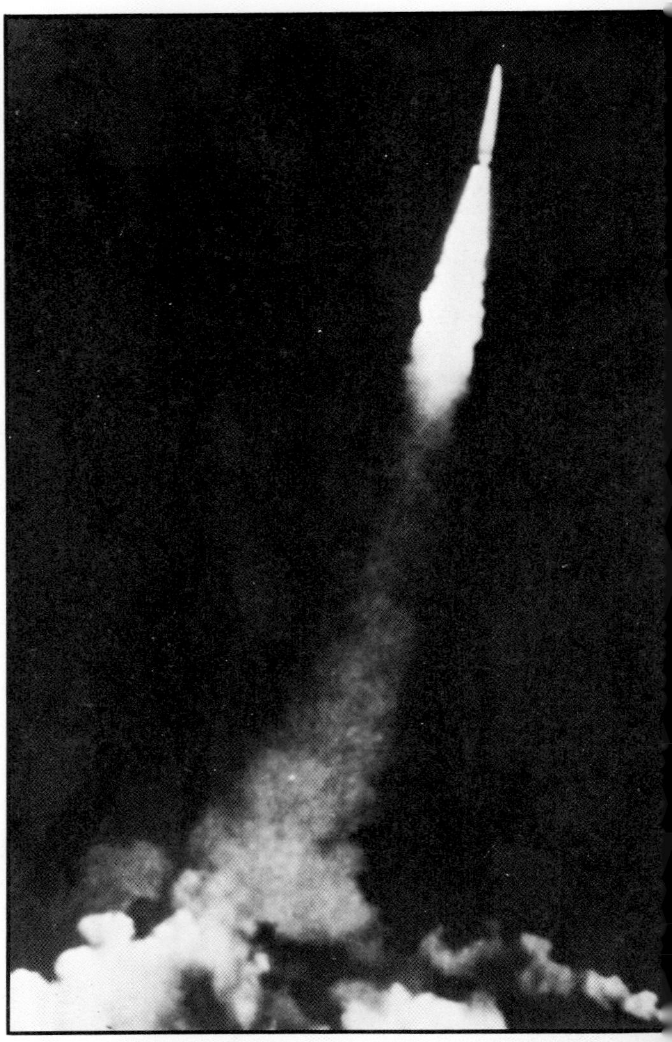

Right: A test launch of an Intermediate Range Ballistic Missile in 1982.

Below: A 'Second Artillery' unit deploys an ICBM during exercises in 1980.

tactics and only attacked when able to catch the UN enemy unaware – melting away into the night when the opposition looked too strong.

By the end of 1950 the UN forces had been driven pell mell southward into South Korea but then they held their ground. Using an enormous concentration of fire power they gradually blasted their way back northward to the 38th Parallel. The massed lightly armed Chinese infantry suffered fearful casualties and simply could not combat the technical superiority of the UN forces despite a fierce resistance.

This, the PLA's first experience of full scale modern conventional war was to have radical repercussions within the Chinese army and on the doctrine of the people's war. Taking advice and military aid from the Soviet Union the Main Forces of the PLA were armed, equipped and trained to fight conventional war in the years following the end of the bitter conflict in Korea in 1953.

India

In the late fifties the frontier area between the PRC and India saw tension building up. There had always been a dispute over the demarcation of the frontier in the Ladakh region of Kashmir and further east in the Assam area of the North East Frontier Agency. The Chinese claimed that India was occupying large tracts of what was legitimately Chinese territory but it may be that the Chinese military build up in the area was meant to serve as a caution to India against any move to support the dissident elements in Tibet where the PLA was used to put down a major rebellion in 1959.

For whatever reason, the situation on the frontier grew gradually more tense and on 20 October 1962 war broke out between the PRC and India as the PLA advanced in force into Ladakh and the North East Frontier Agency (NEFA). Using the massed waves of infantry in the assault so familiar to those who fought in Korea, but now supported by heavy and accurate artillery fire, the PLA penetrated deep into Indian territory. In the Ladakh area the Chinese advanced to occupy the ground they claimed was legally theirs but in the NEFA they pressed forward rapidly for over 100 miles into Indian territory and then suddenly halted. On 21 November 1962 they declared a ceasefire and fell back to the line from where they started the advance on 20 October.

The Indian army, one of the best in Asia, was sadly mauled at the hands of the Chinese and scarcely succeeded in delaying let alone holding the PLA anywhere along the front. The PLA came out of the short sharp war with a considerably enhanced reputation and much boosted morale.

Above: An air defense unit on parade equipped with obsolescent Soviet-designed SA-2 missiles.

Top: Border Guards of the PLA (on foot) warn Soviet troops of a border infringement on the Ussuri River in 1969.

Above right: Chinese Border Guards observe Soviet troops beyond the Ussuri River. The island of Ch'en Pao where the major clash occurred in 1969 can clearly be seen.

Above: Equipment and rations abandoned by the PLA during a skirmish on the Sino-Soviet border.

The Sino-Soviet border

The border between China and Russia in the Far East, between the Russian (now Soviet) Maritime Province and Manchuria, has historically been a source of contention between the two powers long predating the arrival of communism in either country. In essence the Chinese claim that the territory of the Soviet Maritime Province (where the Trans-Siberian railway reaches Vladivostok) is rightfully theirs.

Tension along this section of the frontier demarcated by the Amur and Ussuri Rivers inevitably heightened with the Sino-Soviet rift. For the Chinese it is a means to bring pressure to bear on the Soviet Union as well as a possible opportunity to regain control of what they see as legitimately Chinese territory. The problem for the Soviet Union is real enough. If for instance, they were to concede but one particular small island in the Amur River then the Chinese would have the Trans-Siberian railway, where it runs through the major garrison town of Khabarovsk, within range of their artillery.

Since the early sixties the tension has escalated along the frontier with each side claiming incursions across the border by the other. Indeed, the Chinese claimed in 1969 that troops of the Soviet Union had provoked trouble in the area on no less than 4189 occasions since 1964.

In the Military Regions adjacent to the border it is generally the border troops of the PLA which have dealt with these problems and made their own forays into Russian territory if the Soviet authorities are to be believed.

The biggest flare up on the frontier occurred in 1969. On 2 March a unit of PLA frontier guards ambushed and virtually wiped out a patrol of Soviet troops. The patrol had crossed the frozen Ussuri River to reach the island of Ch'en Pao on the Chinese side of the main stream of the river. The Soviet forces in the region reacted swiftly to the Chinese ambush and severely shelled the Chinese side of the river. The area was the scene of further action in the ensuing weeks and it is believed that battles were fought at regimental level at the time. The Soviet Union augmented its forces all along the frontier and other serious incidents occurred even as far afield as in Sinkiang.

Tension remains high to this day and over the years has been accompanied by a continuous build up of troops. PLA Main Forces are strongly deployed in the Shenyang and Peking Military Regions covering the frontier with a Mongolian People's Republic where several Soviet divisions are deployed and to the south of the Ussuri and Amur Rivers adjacent with the Soviet Maritime Province.

The Sino-Vietnamese War

Throughout the wars against the French, the US and the South Vietnamese China had supported the Hanoi leadership whenever possible but after the death of Ho Chi Minh in 1968 his successors veered increasingly toward seeking support from the Soviet Union rather than the PRC. In consequence the Chinese grew more and more alarmed at the prospect of a strong ally of the Soviet Union to the south.

Their fears seemed to be confirmed when, in November 1978, the Vietnamese signed a treaty of friendship and cooperation with the USSR and in the following month moved into Kampuchea, the PRC's ally. Already the Vietnamese writ virtually ran in Laos and it seemed likely that Hanoi might soon rule the whole of Indo-China and, supported by the Soviet Union, pose an even greater threat.

Relationships between Vietnam and the PRC deteriorated throughout 1978 and the Chinese were probably incensed by the Vietnamese government's treatment of the 'overseas' Chinese – long resident in Vietnam – who were bullied and harried throughout the country.

On 17 February 1979 the PLA launched an attack across the frontier into Vietnam. The problems of the Chinese resident in Vietnam appeared to provide the excuse though there is little doubt that the reason for the attack was to caution both the Soviet Union and Vietnam against actions perceived to pose further threats to the interests of the PRC.

It is estimated that some 150,000 men of the PLA were involved in the war against Vietnam with up to 100,000 actually advancing into Vietnamese territory. The bulk of the forces were found from the armies in the two Military Regions adjacent to the Vietnamese frontier but additional specialist forces were drawn from many other Military Regions, particularly the Wuhan Military Region in central China where large numbers of troops are located to provide reserves for such contingencies.

The advance was extremely slow and the outcome of the battles between the PLA and the Vietnamese forces appears to have been somewhat confused with both sides continually claiming victories. It is however clear that the PLA succeeded in capturing provincial capitals in the border areas which they held on to until in March the Chinese army began to withdraw.

The Chinese move against Vietnam was from the outset announced to be a 'short, sharp lesson' to that country. This may have been true but there is no doubt that they had hoped to force the Vietnamese to withdraw at least some of their forces from Kampuchea and this did not happen.

Above: A Vietnamese artillery unit engaging PLA positions during the Sino-Vietnamese War of 1979. The ammunition crates are clearly marked in Russian.

The Threat

Since its creation in 1949 the foreign policy of the PRC has been marked by a succession of major changes. Initially, when Mao came to power and showed every indication that he wished the PRC to move forward in close cooperation and alliance with the Soviet Union it seemed that the two powers dominating the Euro-Asian land mass posed, or would pose as they developed economically and militarily, a major threat to the interests of the western world. At this point in the fifties, communism appeared to the West to be a united force and the threat, both in Europe and East and Southeast Asia, very real.

Then came the Sino-Soviet dispute. With the PRC and the Soviet Union moving apart the threat to the West diminished considerably though Europe remained alarmed at the Soviet military strength behind the Iron Curtain and the US and the countries of Southeast Asia and Australasia continued to watch and wait for the PRC's next move.

Bereft of Soviet technical support, however, China was scarcely strong enough to move other than in a token way against her neighbors. Meanwhile, internally, great efforts to modernize caused major economic and political disruption with China desperately trying, alone, to resolve her problems. At this time, in the sixties, she continued to support communist insurrections, particularly in Southeast and East Asia. Her support for what were dubbed 'peoples' wars of national liberation' remained for the most part at the propaganda level – there were not enough resources for domestic consumption without contemplating making them available for others. Even these very limited involvements in external affairs left the West fearful of China's intentions however. It seemed the PRC remained bent on engineering the collapse of any non-communist states she could influence.

From the late sixties until the present day much of this has changed. The Chinese have always been realists and their view of the world today has led them to recognize the need for bold changes in strategy. In the first place, they perceive a threat from two sources: the Soviet Union and the United States (also as leader of the western world). Secondly, if they are to defend themselves effectively against possible aggression they must modernize their industry, develop the economy generally and ultimately improve their military capability.

To achieve these aims the PRC has joined the nations of the world in playing the balance of power game, both at superpower level and in regional, more localized terms in East, Southeast and South Asia. Assessing the USSR to be the greatest threat and recognizing the antagonism existing between the Soviet Union and the US, China has moved deliberately closer to the US, the western European powers and Japan hoping as a result, at one and the same time, to caution the Soviet Union against any rash move and to receive technical and economic benefits from the closer association with the western nations.

While playing out this role at the highest level the PRC seeks friendship with states in East, Southeast and South Asia, attempting to establish closer ties with them with the aims of achieving economic benefit and curbing the influence of the Soviet Union in these areas. The PRC also continues with, at least, verbal support for indigenous communist parties attempting their own uprisings even in countries such as Malaysia and the Philippines where she also seeks to maintain friendly relations with the existing non-communist governments. She

Above: A machine gun team of the PLA in action against the Vietnamese in 1979.

Opposite: Vietnamese soldiers surrender to the PLA during the Sino-Vietnamese War.

justifies this apparent anomaly by claiming her support for such movements is natural and emanates not from the government of the PRC but from the CCP. Currently such support is in a very minor key, amounting to little more than sympathy, but clearly if Chinese interests seemed to justify it, it could be stepped up.

Almost all aspects of Chinese foreign policy are geared to the continuance of the Sino-Soviet dispute. If relations between the PRC and the USSR were to improve, and this is not inconceivable if unlikely at the present time, then the threat to western interests would instantly be greatly magnified. In particular, Southeast Asia could readily be expected to come under severe threat and the old fears engendered by the domino theory, the toppling of one state after another into the communist camp, could be realized. Were Singapore and the Straits of Malacca to fall into communist hands then the western world would be placed at a major strategic disadvantage with movement between the Indian Ocean and the Pacific denied to them. The Chinese, feeling theatened by the USSR would, however, surely be highly unlikely to enter alone into a period of 'adventurism' seeking territorial advantage in the east when threatened from the north. Even the attack on Vietnam, limited as it was, caused the PRC enough problems in 1979.

In summary the PRC plays a full role on the international scene, seeks allies against the Soviet Union in all parts of the world and offers only moderate support, almost as an insurance policy, to communist movements everywhere. She is currently too weak to do more but her strength slowly increases, her industrial base expands, her general economy develops and the PLA improves its capabilities. In the future, particularly with a full nuclear capability, the PRC may be expected to begin to call the tune rather than dance to that of other pipers.

Below: A PLA unit assists the civil authorities in a clean up campaign in 1982. As in many communist countries the Chinese military is involved in numerous civil projects.

Left: A propaganda poster celebrates the 55th Anniversary of the PLA.

Below left: The post-Mao era. Deng Xiao-ping votes at the 12th National Congress of the Chinese Communist Party in 1982.

4. EAST AND SOUTHEAST ASIA

Historical background

The spread of communism

Communism spread into East and Southeast Asia in the twenties and initially, the impetus came from the Soviet Union. Lenin created a new Communist International (Comintern) in 1919 with the main aim of building up communist organizations throughout the world. The Far Eastern Bureau of the Comintern was based in Shanghai in China and it was this office that was responsible for promoting communist interests in the region. The main aim of the Comintern in Southeast Asia was to foment insurrection with the intention of weakening the hold of the colonial powers in the area rather than helping to improve the lot of the local peoples. Lenin sought to destroy the capitalist system in Europe and argued that if the Europeans lost their colonies the system would inevitably collapse and 'free' the 'subjugated peoples.'

As a result of careful preparation by Comitern representatives in Southeast Asia the South Seas Communist Party was created early in the twenties and it was from this party that most of the other Southeast Asian parties were to spring. These developments were prepared by the Comintern which meant that control of the various organizations which came into being was firmly in the hands of the Communist Party of the Soviet Union from which body in Moscow came all instructions and support.

However, one particular development from the very early stages was to have serious repercussions for world communism right up to the present day. China has always shown an active interest in Southeast Asia. Already in the middle ages Chinese traders were active in the region and commercial shipping carried trade between the homeland and Southeast Asian ports. Throughout the nineteenth century and the first half of the twentieth a major influx of Chinese immigrants came to settle in the region. They came to work in the European colonies where plantations and other enterprises had been developed and also to continue to trade. By 1973 it was estimated that there were some 15 million Chinese resident in the region. In the period between the two world wars the depressed state of the world economy had a serious effect on the Southeast Asian colonies and many of the 'overseas Chinese,' as they had come to be known, experienced grave hardships. The emissaries of communism arrived at much the same time and thus found many Chinese willing to support them. Communist influence was already gaining ground in China itself and this development spurred many of the overseas Chinese to its support in Southeast Asia. At that time the theories of communism ran counter to the beliefs of most indigenous peoples in the region and as a result communism came to be closely linked with the overseas Chinese whether or not they were closely involved. This Chinese link has taken on considerable importance in recent times, particularly since the onset of the Sino-Soviet dispute.

The impact of communism was not particularly great in Southeast Asia up to the outbreak of World War II. Communist parties were formed throughout the area (in the Dutch East Indies (Indonesia) in 1924, in Malaya, Indo-China and the Philippines in 1930) and campaigns of labor unrest, assassination, and sabotage were initiated though none was on a large scale and all were dealt with by the colonial authorities without great difficulty.

World War II

World War II was, however, to be a turning point and after 1945 events were to take on a different momentum. Until 1939 there had been, apart from communism, little political development in any of the colonial territories except, perhaps, in the Dutch East Indies where nationalists were already active in the thirties. In consequence when the Japanese seized control of the whole region in 1941-42, and the European colonists were so decisively ejected from the area, the only political groups capable of reacting to the new situation were the communist parties. Everywhere throughout Southeast Asia, communists adopted the Comintern line which was anti-fascist, and thus anti-Japanese. Mao Tse-tung took up a similar stance in China and where overseas Chinese were active inside communist parties in Southeast Asia, from then onward they followed Mao's lead as much as that of the Comintern.

Communist opposition to the Japanese occupation was everywhere on a very limited scale, except, perhaps, in Malaya where the Communist Party of Malaya (CPM) established the Malayan People's Anti-Japanese Army (MPAJA). The MPAJA was well organized as an underground force and though it was not capable of being more than an irritant to the Japanese it did serve as the basis on which to build a guerrilla force after the war.

Post World War II Developments

Indo-China. Ho Chi Minh, the communist leader in Indo-China, was a figure of major importance in the world communist movement. He sought in Indo-China to weave together a blend of communism and nationalism so that he could build up a strong opposition to continued

Left: The calm before the storm. Ho Chi Minh confers with the French General Leclerc in Hanoi in 1945.

Below: President of the Republic of (South) Vietnam, Ngo Dinh Diem, greeted in Washington by President Eisenhower in 1957. Diem desperately wanted military assistance from the US but the corrupt nature of his regime weakened his position.

Above: Elated Vietminh soldiers celebrate the shooting down of a French transport aircraft.

French rule there after the war. By 1945 he was strong enough in his home territory in Tongking (now the northern part of Vietnam) to lay claim to govern the territory after the Japanese capitulation. The French, however, returned in force and succeeded after some localized and bitter fighting in re-establishing their authority in Hanoi, the capital of Tongking. Ho withdrew his forces, the Vietminh, to the northeast into the mountains close to the Chinese frontier and from there he and his forceful military commander Vo Nguyen Giap built up their army and prepared to fight the French making use of Mao Tse-tung's guerrilla theories as a basis for their actions.

The French initially held the towns but as Giap's guerrillas became stronger they were able to operate against the French communications making their hold on the towns ever more tenuous. Despite major campaigns conducted against the Vietminh, hoping to bring Giap's forces to a conclusive battle, the French position weakened and their efforts finally foundered following the debacle at Dien Bien Phu in 1954.

From 1949 the Vietminh were sustained in large measure from across the Chinese frontier as a result of Mao's success there. Vietminh forces were granted sanctuary in Chinese territory when pursued by the French and the supply by the Chinese of artillery and other weapons allowed them to build up sufficient strength to defeat the French when the latter played their last card at Dien Bien Phu.

At a conference at Geneva in 1954 the French acknowledged defeat and withdrew from the whole of Indo-China. Laos and Cambodia were given independence on a non-aligned basis and Tongking, Annam and Cochin-China were divided into two: the northern territory becoming the Democratic Republic of Vietnam under Ho Chi Minh and the south the Republic of Vietnam. Paper agreements were signed arranging for the two Vietnams to be united within four years following elections held throughout the two territories.

Vietnam. From the outset it seemed clear that voluntary unification would not take place and the scene was set in which force would be the deciding factor. From 1954 onward the North Vietnamese People's Army (NVPA) strength was built up largely through assistance from the People's Republic of China (PRC) and increasingly from the Soviet Union and it soon threatened the position of the South Vietnamese. The latter's position was made worse by the growth of a communist guerrilla movement, the Viet Cong, within the rural areas of South Vietnam itself. Links were strong between the NVPA and the Viet Cong and the former were almost continuously able to supply the Viet Cong with necessary war materials to fight their campaign by transporting them southward along remote tracks in the frontier area, between Laos and Cambodia with Vietnam, which came collectively to be known as 'the Ho Chi Minh trail.'

After the French collapse the US became increasingly involved in support of the Republic of Vietnam. By the late fifties massive military aid was arriving and thousands of US military advisers were in the country helping to strengthen and increase the effectiveness of the Army of the Republic of Vietnam (ARVN) – the southern army.

The early sixties saw strong attacks both by the Viet Cong and the NVPA and to avoid what appeared to be an inevitable slide toward communist victory in the South the US decided,

Left: Instant and brutal retribution. South Vietnam's Chief of Police shoots a Viet Cong officer captured in Saigon during the Tet offensive in 1968.

Below: Men of the Viet Cong advance to attack a South Vietnamese Army stronghold.

Above: A remote US Army fire support base in Thua Thien Province in 1969. Although such bases were normally sited in mutually-supporting positions, they were a favorite target for Viet Cong attacks.

Right: A US airborne soldier admires the handiwork of a Viet Cong artist during a search and destroy operation in 1967.

in 1965, to commit substantial numbers of American troops to the fighting. Initially, American involvement seemed to wrest the initiative from the communists though it needed, by 1967, some quarter of a million troops to do so. Then, in 1968 the NVPA and the Viet Cong launched a joint attack to coincide with Tet, the Vietnamese New Year.

The Americans and South Vietnamese were caught off balance as the NVPA attacked across the Demilitarized Zone in force and the Viet Cong struck at almost every town in the South. The fighting lasted at a high level for over a month and ended with the communists having incurred severe casualties. However, redeployments of both US and ARVN units to counter attacks from what seemed all directions left many remote regions of the South firmly under the control of the Viet Cong, an advantageous position they were never again to lose.

1969 saw the US decision to disengage from direct military intervention in Vietnam. Negotiations were begun between North and South and the US from then onward and the last US troops left South Vietnam in January 1973. Despite effective and often fierce fighting by the ARVN, slowly but surely the combined forces of the communists forced the initiative away from the South and Saigon, the capital, sur-

rendered unconditionally on 30 April 1975. The whole of Vietnam was finally united under the communists and named the Socialist Republic of Vietnam.

Laos. From 1945 a Laotian communist force, the Pathet Lao, fought against the French until their departure in 1954. Following the achievement of independence in that year there began a struggle between Laotian government forces and the Pathet Lao. The former were supported by the US which tried to swing them over to active opposition to the North Vietnamese whose forces made continual use of the sanctuary of Laotian territory to support their actions against the South. On the other hand, the Pathet Lao were supported throughout by the North Vietnamese until, at last, they achieved victory against the government in 1975. Since then the Lao People's Liberation Army (as the Pathet Lao then became), trained and advised by the Vietnamese People's Army (VPA) has effectively controlled the country which is closely tied to the SRVN.

Cambodia (Kampuchea). Nationalist rather than communist forces resisted the return of the French to Cambodia in 1945 and these same forces led by Prince Sihanouk took the country into independence and neutrality after 1954. Thereafter Sihanouk found himself walking a tightrope balancing between the Americans on the one hand and the Khmer (Cambodian) communist movement and the North Vietnamese on the other. The communists were determined to make use of Cambodian territory along the Vietnamese frontier to sustain the war against the South and the US was determined to limit the damage that this was causing.

Sihanouk tried to curb the use the communists made of Cambodian ground hoping to fend off American pressure but he was unsuccessful and the crisis came at the end of the sixties. In 1970 his government was deposed by Lon Nol, a former prime minister, who moved away from neutralism toward the side of the US. Sihanouk fled to Peking where he actively promoted the interests of those Cambodians antagonistic to Lon Nol and this included the Cambodian communists, the Khmer Rouge.

Lon Nol was unable to prevent the North Vietnamese and Viet Cong from making use of Cambodian territory (his authority scarcely extended outside the capital city, Phnom Penh) and by the end of the sixties the Americans were planning active intervention in Cambodia to weaken the logistic support for the Viet Cong which the Ho Chi Minh trail provided. These efforts were however largely unsuccessful and the forays that did take place caused only temporary breakdowns in the communists' chain of supply which were quickly remedied.

Meanwhile the forces of the Khmer Rouge, under the brutal leadership of a ferocious communist, Pol Pot, mounted a series of vigorous campaigns against the government during which they established their authority almost everywhere in the rural districts of Cambodia. By 1973 the Lon Nol forces had virtually retreated to the capital city and the remainder of the country was in Pol Pot's hands. The Khmer Rouge had by this time, unlike the Laotian communists, severed all links with the North Vietnamese and were receiving aid only from the PRC. Throughout 1974 the Khmer Rouge invested Phnom Penh and in April 1975,

Below: South Vietnamese soldiers wading through a padi field while on the look out for Viet Cong positions in 1963.

Right: Tunku Abdul Rahman, Prime Minister of Malaya confronts Chin Peng, leader of the Malayan Communists in abortive truce talks in 1957. The Malayan emergency went on until 1960 and some fighting continues today.

just as the North Vietnamese marched into Saigon, Lon Nol fled from his capital and the Khmer Rouge entered it. Cambodia was instantly renamed Democratic Kampuchea and Pol Pot commenced a brutal campaign to exterminate all those who had sympathized with causes other than his own.

The Philippines. In World War II almost all Filipino political groupings accepted the Japanese occupation without resistance. The Communist Party of the Philippines (CPP) did not follow suit, however, and in 1942 created an organization designed to appeal to nationalist sentiments – the People's Anti-Japanese Resistance Army – commonly referred to as the Huks from a corruption of an acronym of its full name (Hukbalahap) in the Tagalog language. The Huks were predominantly Filipinos though some overseas Chinese were involved and at their remote headquarters on Luzon island a training team from Mao's PLA was active throughout the war.

The Huks were in fact scarcely active against the Japanese during the war but when Japan capitulated they endeavored somewhat rashly to establish a 'people's government' in the Philippines. However, the returning Americans arrested the leadership and the attempt was soon brought to nothing.

In 1946 there were elections prior to the departure of the Americans. The communists were split in the run up to the election: one section of the party advocated a coalition of left wing parties which would pursue socialist policies (coinciding with Moscow's line at the time) while the other section was for using Mao Tse-tung's method to bring about a 'people's republic.' In the event the election was won by a right wing grouping and the Huks went into hiding to develop their struggle along Maoist lines.

By 1950 they had achieved a great deal of success and were in control of many of the rural areas of Luzon, the main Filipino island. Their achievements were as much due to the excesses of the government, which alienated the peasantry and caused them to look to the communists for support, as it was to their own efforts.

In 1950 Ramon Magsaysay, a Congressman who had earned much respect for his honesty, was appointed Minister of Defense. He was possessed of enormous energy and considerable political insight and he galvanized the army into action. Trained rapidly in the art, it became adept as a counter-insurgency force and, succeeding quickly in rounding up the Huk leadership, the movement began to crumble.

In 1953 Magsaysay was elected President of the Philippines and he worked enthusiastically to win the peasant farmers to his side by ensuring them a fair deal and an end to persecution from commercial interests. Such policies caused a total collapse of the Huks by 1954 and despite successive shifts to the right by succeeding governments no communist movement of strength reappeared in the Philippines for some time. The New People's Army, founded in March 1969, is now active and is estimated to have some 10,000 men involved in operations in various areas, particularly on Mindanao.

Indonesia. The communist movement came out of World War II in a poor state of organization. The immediate post war period saw the Republic of Indonesia formed (1949) out of the Dutch East Indies with scarcely any communist influence apparent. There was at the time much infighting among the leadership of the Indonesian Communist Party (PKI) and as a result no co-ordinated strategy. An attempt to establish communist government in a small area of Java was crushed by nationalist groups supporting Dr Sukarno who was President of Indonesia from 1949 until deposed in 1965.

Although some three million overseas Chinese were resident in Indonesia, almost none belonged to the PKI which was always very much dominated at all levels by indigenous Indonesians. As the years passed, after 1949, Sukarno adopted a more and more benevolent attitude to the PKI which in consequence prospered until by the early sixties its membership touched something near three million. It became very influential throughout the country with many members of the party in government service and in the armed forces.

However, Sukarno's great ambitions for Indonesia were to prove the downfall of the PKI. In the early sixties he adopted a militantly antagonistic policy toward the formation of the Federation of Malaysia which involved the union of Malaya and Singapore with two British Bornean colonies: Sarawak and British North Borneo (now Sabah).

Sukarno pursued a policy of confrontation and began to build up the strength of his army in Borneo along the frontier with Sarawak and Sabah. By 1965 the costs of confrontation, coupled with a seriously declining economy and uneasiness at the President's ever more sympathetic support for the PKI, led to an internal upheaval which resulted in Sukarno being ousted from the Presidency, the complete elimination of the PKI and the establishment of a new and virulently anti-communist military government.

Malaysia. The Communist Party of Malaya (CPM) came out of World War II with a very effective organization and some military experience having created the Malayan People's Anti-Japanese Army during the war. Although it confined its immediate post war activities to developing unrest among labor forces in the towns and on large rubber estates it was not long before its apparent successes prompted it to enter upon an insurgency campaign with the aim of creating a Malayan communist state.

The 'emergency,' as the British authorities called the insurgency period, began in 1948 and lasted until 1960. In the early stages it seemed that the insurgents might have a very good chance of success but the government's response was forceful, its policies were effective and resources were sufficient and the bid for power did not come off.

The CPM may be seen to have failed for a variety of reasons of which perhaps the most important were: the CPM was almost totally comprised of overseas Chinese, and Malays (the bulk of the population) were never sym-

Above: A member of the Communist Malayan Races Liberation Army surrenders to the Malayan security forces in 1950.

Left: Dr Sukarno in 1966, shortly after he was ousted from the Presidency of Indonesia.

Above: Filipino soldiers and police gathered round the bodies of Huk (Communist) guerrillas they have just killed in an ambush.

pathetic to their cause; the British government, early in the emergency, promised independence to Malaya (it got it in 1957) and thereby stole the communists' thunder and lastly, the CPM was not equipped and trained to fight against the sophisticated forces that were ranged against it.

By 1960, when the emergency came to an end, the remnants of the Malayan Races Liberation Army of the CPM had been driven northward across the Malayan border into the jungles on the frontier with Thailand. They were not totally destroyed, indeed, their leader throughout the emergency, Chin Peng, remains at the head of the movement.

The sixties were used by the CPM to regroup and build up their strength on the Thai border and from 1969 they cautiously commenced a new insurgency campaign which continues to the present. It still remains the case, however, that armed and uniformed guerrillas confine their activities for the most part to the northern states in which they can operate most successfully from across the Thai border.

The new insurgency is different in two main respects from that of the 1948-60 period. In the first place there are now large numbers of disaffected Malays within the guerrilla organization which has thus assumed a multiracial composition. This means that communism is more attractive to the poorer elements of both the main races: the Malays and Chinese. On the other hand the organization has since the early seventies been split into factions, each having its own guerrilla forces. This development seriously weakens the impact that communism might otherwise achieve throughout the country.

Insurrection has also been attempted by communist groups in East Malaysia (Sarawak and Sabah) but overall with only limited and occasional success. In East Malaysia as in the peninsula the bulk of the communists are overseas Chinese.

In Sarawak in the fifties, while still a British colony, there were active communist cells in being and these increased in number and greatly enlarged the scope of their activities during the period following the formation of Malaysia from 1962 to 1965. They opposed the creation of Malaysia and thus shared the aims of Dr Sukarno, the Indonesian President, whose policy of confrontation with Malaysia was aimed at destroying the new federation. The incredibly remote jungle areas on the border of Sarawak and Indonesian Borneo (Kalimantan) offered ideal country for the Sarawak communists to establish their secure bases and when pressed by security forces they were able to seek sanctuary across the border in Indonesian territory.

All this came to an end, however, following the coup which ousted Sukarno and established a military government in Indonesia which was and remains staunchly anti-communist. From the mid sixties the Sarawak communist movement went into decline. Mass defections whittled away its numbers until there remained only a hard core of perhaps two hundred men whose activities were greatly circumscribed by the security forces even if they could not be entirely prevented.

Thailand. Unlike other areas of Southeast Asia Thailand was not colonized by a European power. In consequence it did not experience the traumas that affected the newly emerging Southeast Asian states in the post World War II period. Peaceful and comparatively prosperous conditions for the urban and rural populations alike also provided stony ground for the seeds of communist subversion to fall on. Although the Communist Party of Thailand (CPT) had been in being since the early thirties it failed to gain political ground of significance until well into the sixties and it was not until then that Thai governments were faced with effective insurgency activity.

When guerrilla forces did emerge they centered on three areas of the country: in the north adjacent to the Burmese border; in the northeast along the frontier with Laos and Kampuchea (formerly Cambodia) and in the south in the jungles just inside the frontier with Peninsular Malaya. The CPT has never succeeded in coordinating its guerrilla efforts in the three areas and as a result has been far less successful than might have been the case.

The remoteness of the three regions one from another provides one explanation for the lack of combined effort but there is another more important factor. While the Thai people form the majority of the population of Thailand there are also large numbers of ethnic minorities: a consequence mainly of the shifting frontiers and ease of migration in the area that have been typical of the last 500 years. Overseas Chinese have been involved in the CPT on a small scale, as have some disaffected Thais, but in the northern insurgency the Meo and Karen peoples provide most recruits, while in the northeast Thais of Laotian origin are involved, and in the south Malays.

In the north the insurgency campaign which broke out in 1967 resulted as much from the bitterness of the Meo people at what they saw as racial discrimination on the part of central government as from communist sources. It was rather a case of the CPT making use of the Meo for their own ends.

1965 saw the beginning of the northeastern insurrection which from the outset has been the most dangerous for the government. With the Pathet Lao communist movement in Laos ever more successful and also receiving support from China and Vietnam the Thai security forces were rightly fearful of considerable external support for this campaign. Initially, while they conducted a campaign of assassination, intimidation and propaganda, the guerrillas in the northeast won considerable support but it was not until after 1975, when Laos fell wholly into the communist orbit and was effectively a satellite of a united Vietnam, that the insurrection posed a major threat to the Thai government.

Government has only paid limited attention to the communist campaign in the south. It has cooperated to an extent with the Malaysian government to weaken support for the guerrillas on their joint frontier but, with some justification, it has always seen the problem as primarily one for Malaysia.

North Korea. The Korean peninsula became a colony of Japan in 1910 and remained a Japanese possession until the end of World War II. During this period the Koreans were kept under fierce subjugation with no real chance of giving voice to political opinions. A Korean Communist Party was formed in 1925 but strict security measures forced it underground and it was, as a result, unable to exert much influence in the peninsula. Korean communists seeking to play a more active part in promoting their ideals managed to do so more effectively outside their own country. Many of them joined the Chinese communists in the thirties to fight against the Kuomintang and others sought sanctuary in the Soviet Far East and endeavored from there to build up a partisan force to fight the Japanese.

At the end of World War II Soviet forces, with Allied agreement, occupied the northern part of Korea down to the 38th Parallel: US forces occupied the southern part of the country. It was agreed between the Allies that Korea should become a single sovereign state following elections throughout the country, but this never happened.

The Russians were quick to organize their zone of Korea after their own system both in civil and military terms. In October 1945 they installed Kim Il Sung as General Secretary of the Korean Workers Party which was comprised of communists who had been under Soviet influence and protection during the war. Other Korean communists who had been in China with Mao's armies were assimilated into the party in August 1946 and as a united group, heavily under Soviet influence they proclaimed the Democratic Republic of Korea in September 1948. The Republic of Korea had already, in July 1948, been established in southern Korea.

Both republics claimed the right to govern the whole peninsula and were highly antagonistic toward each other. Such an unstable situation could not last and in June 1950 the highly efficient and Soviet equipped North Korean army marched into South Korea, thus beginning the Korean War which lasted until 1953.

United Nations forces were quickly drawn in to save the South Koreans and then, when the North looked ready to fall, and with an eye to preventing any UN incursion into Manchuria, the 'Chinese People's Volunteers' were launched into battle in their thousands. The UN forces were driven back and the front finally

Right: Kim Il Sung, President of the Democratic Republic of Korea (North Korea) since it came into being in 1948.

Second right: North Korean troops mounted in Soviet T-34 tanks parade in Pyongyang.

Below: The frontier between North and South Korea remains tense. Here North Korean soldiers keep watch on the South.

stabilized approximately on the 38th Parallel and after a bitter period of some two years of static fighting a truce was signed which brought hostilities to an end.

No peace treaty has ever been signed and up to the present the peninsula remains divided, with Kim Il Sung still in power to the north. Throughout the years since he came to power Kim Il Sung has developed his state along socialist lines but with additional elements introduced by him which fit, as he sees it, Korean conditions. He has remained consistently antagonistic to the South although there have been occasions when dialogue between the two Koreas might have been possible.

In foreign policy he has been a realist and sought as far as has been practicable to pursue a neutral policy between the Soviet Union and the PRC. Inevitably this has been extremely difficult since the Sino–Soviet rift. Neither the Chinese nor the Russians have seen fit to attempt to diminish each other's influence in Korea. They both recognize the critical strategic importance that the other attaches to the Korean peninsula and neither could afford to accept a Korean state completely under the thumb of the other.

Sino-Soviet rivalry in Southeast Asia

The Soviet Union was responsible for introducing communism into Southeast Asia. Using the Comintern Far Eastern Bureau in Shanghai as its agent it insinuated its representatives into China and then into the Southeast Asian territories under colonial rule in the twenties and thirties. In the initial stages of penetration the Soviet Union was thus most influential in the Far East. But other factors were to come into play which were to act to curb the Russians' influence and enhance that of the Chinese.

Not all but many of the communist parties found their recruits, as we have seen, from the

Above: Women soldiers of the NKPA train to use anti-aircraft heavy machine gun equipment.

overseas Chinese communities long resident there. They have, despite living in adopted lands, continued to retain close links with China and hardly surprisingly they look to the successes of communism in the PRC as their example rather than further afield to a distant and alien Moscow.

Nonetheless, Russian communist influence came first and remained comparatively strong until after Mao's final success in China in 1949. It was the voice of Moscow that prevailed on the communist parties of Southeast Asia to rise, for the most part prematurely, in 1948. After World War II the Soviet Union established a new organization to direct international communism. It replaced the Comintern, which ceased to function during the war, and was known as the Communist Information Bureau (Cominform). At its inaugural meeting in Poland in 1947 Zhdanov, one of Stalin's right hand men, called on the 'oppressed colonial peoples' to rise and expel the Europeans. There followed, in Calcutta, in February 1948 a meeting of communist front youth organizations where the original message from Zhdanov was spelled out in more detail as it affected the communist parties of Southeast Asia. No fewer than four parties responded to the invitation: in Burma, Indonesia, the Philippines and Malaya, though all but the last failed quite quickly in their attempts.

After 1949 the PRC became ever more influential. In Malaya in particular where the CPM was made up almost completely of Chinese, the Peking influence was to grow from the outset of the emergency. But it was closer to home where the Chinese communists were to establish a major influence by providing not only propaganda but also effective military support. Indo-China shares a long frontier with China and as soon as Mao's government controlled this frontier he was able to assist the Indo-Chinese communists and nationalists in their war to rid the territory of the French. Sanctuaries were offered to Ho Chi Minh's and Giap's men when under threat from French forces, and military equipment and weapons were provided to the Vietminh which probably were absolutely essential to ensure their victory. Certainly, it was field and anti-aircraft artillery from China which caused the final French collapse at Dien Bien Phu.

Until the Sino-Soviet dispute broke out it can therefore be said that the Chinese moved, following World War II, to a position where their views and influence in Southeast Asia began to predominate. After the rift yet other factors were to come into play.

In the first place the Soviet Union had begun to develop its capacity to play a physical rather than a propaganda role throughout the world. It could act more effectively in the Far East to counter the growing Chinese influence, most particularly because it had more and better military hardware to offer communist parties engaged in insurgency.

The PRC and the Soviet Union have been, since the sixties, competing strongly with each other to win support in the region. Both have been quick to make use of national and racial antagonisms that have existed in the area for centuries.

The Vietnamese peoples, the Tongkingese, the Annamites and the Cochinese have historically been under, or fearful of, Chinese domination. It is thus not surprising that, since the chances of Soviet aid have improved, the leaders in Hanoi have moved away from China and closer to the Soviet Union as an insurance policy against possible moves by their large northern neighbor.

The Khmer peoples have, in turn, been fearful of being swallowed up by their bigger neighbors, the Vietnamese, and thus they have

Right: Today, US and North Korean soldiers still come face to face at Panmunjom where the Korean Armistice was signed in 1953.

turned toward the PRC for support which the Chinese have been only too pleased to offer, seeing the relationship as helping to moderate the growing influence of the Soviet Union in Vietnam.

In Thailand, the Thai Communist Party has found itself in a major dilemma. Taking aid and propaganda support equally over the years from the Soviet Union and the PRC and thus Vietnam and Kampuchea, it today fears more and more the influence stemming from Vietnam following the growing Vietnamese control exercised in both Laos and Kampuchea since 1975. Its dilemma is yet to be resolved and in consequence the TCP has tended to become factionalized and thus less effective.

Another interesting development, which may have been influenced by the Sino-Soviet dispute, has been the splitting up of the Malayan Communist Party. Internal strife, centered on disagreements over policy, developed in the early seventies and the MCP splintered. Part became the Malayan Communist Party Revolutionary Faction (MCP-RF) and another the Malayan Communist Party – Marxist-Leninist (MCP-ML) thus leaving a rump MCP. The two new factions were not prepared to continue with the 'long war' in the rural areas strategy, the MCP view, based on Mao's teachings, and opted for a stronger line within an urban setting (seeking urban support suggests a tendency to move toward the Soviet viewpoint).

There has been another interesting development in Malaya which reflects the Sino-Soviet dispute. During the emergency there from 1948 to 1960 virtually all the communist forces were Chinese. Since then considerable numbers of Malays have rallied to the communist cause and joined the guerrillas in the jungles on the Thai border. There is evidence to suggest that the Soviet Union is exerting considerable pressure to influence the Malay communists, hoping in this way to break the monopoly of influence of the PRC over the Chinese majority within the movement.

The picture then is one of unremitting competition between Russia and China throughout the area each hoping to win, at the expense of the other, a stronger position for itself.

Communist forces in East and Southeast Asia

North Korea

The armed forces of the Democratic People's Republic of Korea (North Korea) are 784,000 strong and are backed by other bodies such as the border guards (38,000) and a workers and farmers militia (760,000). Of the armed forces proper the army is by far the largest at 700,000 with the navy at 33,000 and the air force at 51,000.

All three services are part of a unified force under the authority of a single chief of staff. However, at the highest level, the forces are controlled by the Military Affairs Committee of the party (the Korean Workers Party). The Chairman of this committee is Kim Il Sung who is also President of the Republic, General Secretary of the Party and Supreme Commander of the Armed Forces with the rank of Marshal.

The Military Affairs Committee exercises its control on a day to day basis through the Ministry of National Defense in which the Chief of Staff is a vice-minister. He is responsible for all training and operational matters while a second vice-minister oversees political affairs and a third, administration and logistic services.

The army is divided into nine corps under the command of which there are 35 infantry divisions, two armored divisions and three motorized infantry divisions. There are also five independent armored and four independent infantry brigades plus two independent tank and five independent infantry regiments.

While all divisions have their own integrated artillery and engineer support there are also a number of artillery regiments, missile regiments and engineering (river crossing) regiments.

A further major component of the army is made up of the special forces which are organized under a corps headquarters and comprise 20 brigades (of which three are trained in amphibious warfare) and an airborne element.

The navy is based at the ports of Wonsan and Nampo and comprises 19 submarines of Soviet, Chinese and Korean manufacture and four frigates. It also has a fleet of over 400 attack

and patrol craft armed variously with missiles, guns and torpedoes. There is a limited number of landing craft capable of supporting only the smallest scale of amphibious landing.

The air force possesses some 700 combat aircraft comprising three light bomber squadrons, 13 ground attack fighter squadrons and 12 inteceptor fighter squadrons. The transport fleet numbers approximately 225 planes and there are also about 40 helicopters in service. The air force controls a number of surface-to-air missile brigades which are deployed on approximately 40 sites throughout the country.

The role of the armed forces is seen as the defense of the homeland and the most likely enemy is seen to be the Republic of Korea to the south. The major concentration of ground forces is aligned along the frontier with South Korea and naval and air forces are located similarly to counter any possible move from the South. Such a deployment is, of course, not only suitable for defense but also for an offensive move and it should not be forgotten that a second role for the armed forces is declared to be preparedness at all times to intervene in the South should revolution occur there.

North Korea remains reliant for the supply of much of its military equipment on the Soviet Union and the PRC and recognition of this fact leads it to pursue friendly relations with both its neighbors. With an increasing domestic industrial capacity it is trying to lessen its dependence on other states by building up its own arms industry. It attempts to retain a rough parity of weapons and forces with those of the South and there is no doubt that it is well enough trained and equipped to defend itself more than adequately at the present time. Were it to wish to mount an attack against South Korea it would first need to be assured of military aid either from the Soviet Union or the PRC (or both?) but neither of the two big powers would be likely to support such aims, not seeing any benefit to their interests likely to come from a new Korean War at the present time.

Top: 1976 saw increased tension on the 38th Parallel. Here US troops are seen under attack from North Koreans in the Panmunjom jointly guarded area. Two US officers were killed in this incident.

Above: The Soviet Union continues to woo the North Koreans. Former President Chernenko holds talks with Kim Il Sung. At right is Soviet Foreign Minister Gromyko.

Above: A Chinese aircraft is prepared for battle in the 1979 war against Vietnam.

Above right: North Vietnamese Russian-built MiG-21s ready for combat against the South in 1968.

Right: Vietnamese soldiers man a heavy machine gun during the war against China in 1979.

Vietnam

The armed forces of Vietnam are extremely large, comprising some 1,029,000 men. The vast preponderance are in the army with only approximately 4000 in the navy and 25,000 in the air force.

Supreme command of all the forces rests with the President of the Republic to whom the Minister of Defense and the Commander-in-Chief are answerable. Direction of the forces on a day to day basis is via staff channels which embrace all command, political and administrative functions.

It is believed that Vietnam is divided into seven military regions which provide administrative support to forces in the area. The ground forces are organized in armies which, in turn, are comprised of some 58 divisions, all but one of which are infantry divisions, of three infantry regiments (each of three infantry battalions), an artillery regiment, a tank battalion and other supporting elements. The remaining division is an armored division with three tank regiments, an infantry regiment and supporting arms and services.

Independent of these basic formations, and available to support any grouping of them, there are two marine divisions, seven engineer divisions, five independent field artillery brigades, four anti-aircraft artillery brigades and six independent armored regiments.

The small navy is organized to provide coastal protection and control and river patrolling operations while the air force with a total of 470 aircraft (not all serviceable) is geared to provide limited bombing (one light bomber squadron) and interceptor (12 squadrons) capabilities. There is a major ground attack fighter ability with some 20 squadrons to support the ground forces. The air force also offers limited transport and helicopter services.

In addition to the full time embodied military forces, Vietnam has significant para-military forces with various functions, including frontier and coastal guard units with a manpower of approximately 70,000. There is also a regional armed militia which embraces all persons of military age and includes many who have served in the armed forces. The militia provides the only form of reserve and numbers some 1,500,000.

The overall responsibility of the armed forces is to defend the state and party from any internal or external threat. In pursuit of long established policies of exercising influence throughout Indo-China, some forces, at least 45,000 men, are stationed, by agreement, in Laos. There are also considerable elements of the armed forces currently deployed in Kampuchea (approximately 170,000 in 22 divisions) though the force levels have dropped slightly since just after the invasion of 1978 when it was estimated that up to 200,000 troops were involved.

Internally, the armed forces of Vietnam are, like communist forces worldwide, frequently involved in tasks in support of agricultural and industrial enterprises as well as the construction and maintenance of communications. There are estimated to be about fifteen economic construction divisions established for these purposes.

The Vietnamese forces, if only because of their size, are capable of exerting considerable regional pressure. While not proving successful in overcoming resistance quickly in Kampuchea they appeared able to match the Chinese PLA in the short war of 1979.

Because of its historical background, the army posseses an extraordinarily varied inventory of weapons, vehicles and equipment of Russian, Chinese and American manufacture.

012
5006
4326

Above: An anti-aircraft gun crew of the Laotian Army.

Right: Laotian infantry soldiers on patrol armed with Soviet weapons, principally the ubiquitous AK-47.

It has some 1500 tanks including Soviet T-34, T-54/55 and T-62, Chinese T-59 and US M-48; approximately 450 amphibious tanks, including Soviet PT-76 and Chinese T-60/63. Its artillery, heavy, field and anti-aircraft, includes an even more bewildering array from the same sources.

The inventory suggests on first consideration a force of considerable power but severe reservations must exist regarding the supply of ammunition and spare parts for equipments of both Chinese and American origin. Similarly, the enormous size of the army suggests a force capable of successful combat, at least in regional terms, but while difficult to gauge it is to be expected that grave economic stringency added to years of unabated conflict have taken their toll on the morale of the forces at home as well as those still fighting in Kampuchea.

Laos

Large in territory but small in population, Laos boasts armed forces of only some 48,700 men: rather less than the force of Vietnamese currently stationed in the country. The army predominates, accounting for 46,000 of the total under arms and leaving the navy with 1700 and the air force with 1000 men.

The ground forces are organized on a regional basis with the country divided, it is believed, into five military regions. There is no military hierarchy above battalion level and it is currently assessed that the Laotian army comprises one armored battalion, probably located in the capital, Vientiane, and seventy infantry battalions distributed among the military regions. There are also four artillery battalions and four anti-aircraft artillery battalions similarly located throughout the country in support of the infantry.

In land-locked Laos the navy's task centers on patrolling the main rivers and for this purpose it has something in excess of 30 river patrol craft plus a small number of transport vessels. The air force possesses one squadron of

MiG-21 interceptor aircraft, one squadron of counter-insurgency aircraft, one helicopter squadron and two squadrons of transport aircraft. All squadrons save the interceptors include a mixture of aircraft of which some are of American origin and it is doubtful whether all these latter remain serviceable through lack of spare parts and technical expertise.

The role of the army remains one largely of counter insurgency. Despite having been created in 1975 the People's Democratic Republic of Laos remains confronted with an insurgency campaign which is succoured from both Thailand and the PRC and the deployment of forces, and those from Vietnam, is aimed at reducing the effect of this insurgency.

Above: A 13 year old 'soldier' of the Free Khmer coalition. The photo was taken a year after Vietnam invaded Kampuchea.

Right: Pol Pot, brutal leader of the Khmer Rouge who seized power in Kampuchea in 1975, pictured here in the Cambodian jungle in 1979.

Kampuchea

Kampuchea is in turmoil with two opposing factions claiming to be the legitimate government. Both maintain military forces in the field but one has command over the largest part of the country and controls the capital, Phnom Penh.

In 1975, as North Vietnam finally overcame the South to unite the country under communist rule, the communist movement in Kampuchea, the Khmer Rouge, under its leader Pol Pot also seized power in Kampuchea. At this time the Vietnamese communists were on friendly terms with the Khmer Rouge, albeit somewhat cautious over the latter's close ties with the PRC. Over the ensuing three years Pol Pot proceeded to 'communize' Kampuchea with brutal ferocity, eliminating all opposition to his newly created state of Democratic Kampuchea. The three years saw a marked deterioration in relations between the Khmer Rouge and the Vietnamese, partly because of the tensions between the PRC, supporting the Khmer Rouge and the Soviet Union, supporting Vietnam, but also because the many ethnic Vietnamese long resident in Kampuchea were among Pol Pot's victims and, finally, because of the horrors of Pol Pot's brand of communism as seen in action.

In 1978 the Vietnamese marched into Kampuchea to crush the Khmer Rouge and to support the establishment of a Kampuchean government more sympathetic to their views. The Vietnamese-installed state, led by a

Left: Prince Norodom Sihanouk (center), leader of the Kampuchean coalition against the Vietnamese-sponsored government inspects his soldiers in 1983.

Below: Khmer Rouge soldiers patrol the deserted streets of Phnom Penh after Pol Pot, their leader, had driven the towr speople into the countryside in 1975.

Above: Soldiers of Heng Samrin's pro-Vietnamese Kampuchean army on parade in Phnom Penh.

Right: Anti-aircraft gun crews of the pro-Vietnamese Kampuchean forces.

Khmer, Heng Samrin, is called the People's Republic of Kampuchea. Following quickly after the invasion, the Khmer Rouge were driven westward until they now remain isolated in a number of districts adjacent to the Thai frontier.

Other, non-communist factions in Kampuchea have rallied against the Heng Samrin government, largely because they fear a complete Vietnamese take over and see in the Heng Samrin government nothing but a puppet regime of the Vietnamese. These factions, with forces loyal to them, joined with the Khmer Rouge in 1982 to form a Coalition Government of Democratic Kampuchea. Pol Pot has stepped down from the leadership of the Khmer Rouge (it being felt his name was too closely associated with the early atrocities) and his replacement Khieu Samphan leads the coalition together with Prince Norodom Sihanouk, closely involved in Khmer politics since before the French left in 1954, and Son Sann, leader of the Khmer People's National Liberation Front.

Detailed information on the composition of the armies of the Heng Samrin regime is very difficult to obtain but it is estimated that the army of the People's Republic of Kampuchea is approximately 20-30,000 strong. It is understood that these troops are formed into four weak infantry divisions, each of three brigades. There are also a number, perhaps 50, of smaller independent units including reconnaissance and artillery groups.

The forces loyal to the Coalition Government of Democratic Kampuchea are said to number approximately 30-40,000 but it is virtually impossible to verify the figure.

Thailand

Since the late seventies the Thai Communist Party (TCP) has been faced with grave problems, largely stemming from the Vietnamese invasion of Kampuchea in 1978. The TCP has always leaned toward Peking in the Sino-Soviet dispute though receiving support, verbal and material, from both the PRC and the Soviet Union oriented Vietnamese. While Pol Pot's Peking-supported Khmer Rouge government prevailed in Phnom Penh the Chinese link was dominant but when the Vietnamese forces in Kampuchea arrived on the Thai border major clashes broke out in the TCP between the pro-Peking and pro-Moscow elements.

In consequence the force levels of the TCP's guerrilla movement began to drop following defections from the cause. It is currently believed that the total insurgency forces of the TCP number approximately 8-10,000 with 4000 in the northeast, 3000 in the south and perhaps 2000 in the north.

These groups are armed only with small arms and possess the capability to carry out demolitions and ambushes on road or rail communications. They suffer from grave shortages of military equipment, arms and ammunition because supplies from Vietnamese and Kampuchean sources have dried up and, more important, because the PRC, always the main supplier, has, in its pursuit of friendly relations with the Thai government, reduced its support to little more than muted propaganda.

There continues to be a steady trickle of defections from the TCP and this suggests strongly that the effectiveness of the TCP and its forces is at present low.

Below: A soldier of Heng Samrin's pro-Vietnamese army under instruction.

Malaysia

As far as can be ascertained the guerrilla forces of the Malayan Communist Party (MCP), known as the Malayan National Liberation Army, comprise some 2-3000 men. Well over half of the force is made up of overseas Chinese but the Malay element is known to be increasing.

Until the split within the party organization in the early seventies it was known that the activities of the guerrillas were controlled and coordinated by the Central Committee of the MCP which issued orders to active units through the North Malayan Bureau (in the northern states of the peninsula) and the South Line Organization (covering the southern states).

The active units comprised three: the 8th, 10th and 12th Regiments, of which the 8th and 12th were largely composed of Chinese and the 10th of Malays. There is also an underground network of cells known, as it was during the emergency of 1948-60, as the Min Yuen, which provides intelligence and supplies food and medical supplies to the guerrillas.

Since the MCP-RF and MCP-ML splinter groups have broken away the organization has been somewhat difficult to confirm. It appears that each of these groups attracted something in the order of 500 guerrillas to its cause and both have established underground supporting movements. Both the splinter groups and the original MCP organization have their headquarters located in remote jungle areas just inside the Thai border.

All three groups of guerrillas are armed only with small arms, including some machine guns. They are capable of obtaining some supplies of explosives with which to set up ambushes of security forces.

On current estimates the capability of the communist guerrilla movement in Malaya must be rated as weak. In the first place the splintering of the organization, while initially engendering considerable competition, can only lead to a lessening of the overall effect of their cause in Malaya and, secondly, as in Thailand, active support from the PRC is very much in abeyance at the present time owing to China's pursuit of friendly relations with the Malaysian government while she struggles to counter-balance the Soviet Union's increasing influence in the region.

The communist insurgency forces in Eastern Malaysia (Sarawak and Sabah) have continued to decline in numbers in the last few years to a present estimated level of some 150-200 active guerrillas. The effectiveness of the organization is curtailed by the almost complete lack of external support.

The Threat

The extent of the communist threat to western interests in East and Southeast Asia cannot be estimated without taking into consideration the capabilities and intentions of the Soviet Union and the People's Republic of China in the region.

If the Russians and the Chinese could heal the breach betweem them it is clear that the threat would be vastly increased. The Soviet Union has the ability to operate with naval and air forces in the region and the Chinese PLA is a powerful force which could be utilized to support such Soviet activities. Close cooperation between the USSR and the PRC would probably also place the larger forces of Vietnam and North Korea at their joint disposal. And, further, the weakness of the communist guerrilla forces in the Southeast Asian states, particularly Thailand and Malaysia, would to a great extent be remedied as pro-Peking and pro-Moscow factions would be reunited and, more important, supplies of weapons, ammunition and other warlike stores would, presumably, be once again available.

In such a situation the threat to the West would be extremely serious in both East and Southeast Asia. South Korea and thus Japan would be directly threatened in the east and in the southeast the delicate balance in Thailand might with ease be tilted in favor of communism, thus opening the way southward through Malaya to Singapore. Then the vital sea lanes between the Indian ocean and the Pacific would be cut.

While the two great communist powers remain at loggerheads, however, the threat to western interests is much diluted, though the constant endeavors by both sides to improve their own positions at the expense of the other could in the longer term lead to increased tensions in the region which might generate their own impetus and drag the area into a state of armed confrontation inimical to western interests.

The North Korean government of Kim Il Sung is currently pursuing a comparatively moderate line in its relations with South Korea but this has happened before when friendly gestures have suddenly given way to acts of open hostility – something that can happen in a virtual dictatorship. There remains no doubt at all that, whatever his present policies, Kim Il Sung would, if he thought it possible, wish to extend his authority to the southern part of Korea. Such an endeavor, if not by way of open warfare, as in 1950, then by a gradual escalation of tension, might lead to such a destabilization of the fragile Korean balance as to threaten the West severely.

In the regional context also another possible threat to stability needs to be considered. The Vietnamese have made no secret of their desire to dominate the remainder of Indo-China, that is Laos and Kampuchea. Already Laos is effectively within their control and Kampuchea is well on the way to sharing the same fate. However, all is not decided and there is staunch opposition to the Vietnamese take-over bid. The situation is very fluid, particularly in western Kampuchea, where the forces of the alternative coalition government continue to hold out. When pressed they seek sanctuary across the border in Thailand and when pressing them the Vietnamese forces are, as they have demonstrated, not particularly averse to engaging in 'hot pursuit' into Thai territory.

The longer this war in Kampuchea goes on the more desperate will the Vietnamese be to conclude it and the more likely they will be to ignore the sensibilities of the Thais. The frontier is not easily defined and runs through remote regions away from population centers and, further, minorities of Thais and Khmers have moved over the border in the last few years to seek a respite from the continuous fighting. Their political allegiance is never certain and their refugee camps create major problems for the Thai government.

Considerable tension in the area is the result and the Thai government is handling the situation with extreme caution but there is no doubt that any extravagant move by the Vietnamese, brought about through desperation, could lead to a further grave escalation of tension. With the western world committed to assisting Thailand and for that matter at least morally to the rest of Southeast Asia the West's interests could be placed in serious jeopardy.

Opposite, main picture: Thai Army infantry soldiers on a training exercise.

Opposite, inset: Guerrilla soldiers of the Communist Party of Thailand photographed in the jungles of southern Thailand in 1978.

5. THE MUSLIM WORLD – AIMING FOR THE GULF?

When the paratroops of the Soviet 105th Guards Airborne Division landed at Bagram Airfield to the north of Kabul, Afghanistan, just before Chistmas Eve in 1979, it seemed to many commentators that the USSR was about to make a reality of a putative age-old dream: access to the Gulf. It is often held that, since the days of Tsar Peter the Great in the seventeenth century, Russia has sought an ice-free warm-water port through which trade could pass all the year round, and which would act as a supply artery for its bases in central Asia, and the chosen place for this facility has been the Gulf. Whether there be substance to this or not, it does pose two questions of present-day relevance: does the USSR still need such a port, and how does the Gulf feature in the intentions of the USSR?

Historical background

By the beginning of the twentieth century all of Muslim central Asia, with the single exception of eastern Turkestan (*Xinjiang*, the New Dominion) which was part of the Manchu Empire of China, was in Russian hands. Russia also exerted a 'sphere of influence' over northern Persia. In 1911, Russian troops crushed the socialist Tabriz Revolt in that Persian city. Britain fought three Afghan Wars, largely to eliminate Russian influence from the country which was considered to provide a *cordon sanitaire* for the security of India. British imperial policy was also aimed at shoring up the Asiatic provinces of the Ottoman Empire, and dominating the Gulf and southern Persia, in order to thwart perceived Russian ambitions toward India. Britain was also anxious to prevent the Russians controlling Istanbul ('Constantinople') and the Bosphorus, which would have allowed them free access to the eastern Mediterranean – thereby posing a threat to shipping from the Black Sea ports.

The British attitude toward Russia's expansion seems to have been conditioned by considerable misunderstanding, or ignorance (both geographical and otherwise). Russian expansion in Central Asia appeared to menace British interests, and so British imperial policy was to forestall Russian domination on the periphery of northern India and to create buffer states in which British-Indian influence would predominate. Russia, too, had its subscribers to the forward school. As the Central Asian states fell to the Russian Empire, it came to seem to both British and Russian alike that Afghanistan had to be within the sphere of influence of one or the other.

Ironically, it was the success of the British

Right: An Afghan soldier guarding Aero L-39C trainer-COIN aircraft of the Afghan Air Force. The plane on the right is camouflaged for a light attack role. Two UV-16-57 rocket pods seem to be on the ground under the tailplane.

Right: The Libyan leader Colonel Mu'ammar al-Qaddafi photographed at Moscow airport with Soviet President Leonid Brezhnev when he arrived for a visit in April 1981.

expansion into the Punjab and the channels of contact that this opened up to the Muslim world that helped to break the cultural isolation of Muslim central Asia under which it had languished since the Russians had conquered the Khanate of Astrakhan in 1554 at the time of Tsar Ivan the Terrible. The Russian conquest of the lower Volga had cut the overland route to the Ottoman Empire, and the mainstream of Islamic thought, and also adversely affected the Central Asian economies.

Initially Russian expansion at the time of Ivan had been accepted by the Kazan and Volga Tartars as he was seen as having a rightful dynastic claim to leadership of the *Aqordu* (the 'White' – or western – Horde) and the *Ulus* of Batu. Strangely, considering how tyrannical Tsar Ivan was to others, his policy toward the Muslims was tolerant and enlightened. Proselytizing of the Muslims by the Orthodox Church was forbidden and the Tartar aristocracy was accepted into the ranks of the Russian nobility. Although some of his successors also followed this policy, most of the Tsars were quite intolerant of Islam and encouraged the Church to missionary endeavors. This was sometimes accompanied by deportations, and led to a mounting anti-Russian and anti-Slav feeling among the Central Asians which has endured and intensified to this day. Ivan's liberal vision of a Russian Empire belonging to both Christian Europe and Muslim Asia did not outlast his death in 1584. Thereafter the Muscovy Tsars discovered a link with Byzantium and the mantle of 'New Rome.' It became a policy to spread Russian, Orthodox culture. This led to Russia's self-appointed role as the protector of all Orthodox Christians in the Ottoman Empire in the eighteenth and nineteenth centuries. The logical consequence of this was Russia's desire to conquer Constantinople and re-establish the place of the city in the new 'New Rome.'

In this expansion to the east and south – Russia's *Drang nach Osten* – the cultural contacts were for the most part the opposite of those on America's westward moving frontier. The 'high civilization' was that of the conquered Central Asians, and the '*Rus*' were the barbarians. The legacy of this is felt to this day, with the Russians having a peculiar inferiority complex toward their *chernozhopye* ('black bottoms': a derogatory epithet analogous to 'niggers') and their parent culture.

By the end of the nineteenth century, Russia had acquired all the Central Asian lands to the north of Ottoman Turkey, Qajar Persia, and Afghanistan. Her drive toward Istanbul had been arrested to British satisfaction, but this remained a Russian aim. But cultural influences from the south on her Central Asian peoples continued to worry the Tsarist government. Two field armies had been tied down in the Caucasus fighting the Daghestani supporters of a revivalist movement which had begun in British India (or Delhi under British influence) from where it had spread eventually via the Ottoman Empire to the Caucasus and other parts of the Russian dominions. Such influences were unwelcome, to say the least, to the Russians as they made the task of 'pacification' so much more difficult.

Seen from St Petersburg, the southern border represented not so much a leaky bucket as a sieve through which people and ideas were flowing easily in both directions, not only exacerbating the problems of controlling a bellicose and alien colonial empire but also providing a conduit through which new – and unwelcome – liberal ideas from western Europe were reaching members of the Russian intelligentsia. It is interesting that both Kerensky and Vladimir Ilyich Ulyanov ('Lenin') were

Left: President Gamal Abdel Nasser of Egypt and his deputy and successor Anwar Sadat talking with Soviet leaders Brezhnev, Kosygin, and Podgorny at the Kremlin in July 1968.

born in Central Asia, and Joseph Vissarionovich Dzhugashvili ('Stalin') was a Georgian – next door neighbors to the Daghestanis.

The latter half of the nineteenth century and the early part of the twentieth were also the heydays of two separate, but connected, 'Pan-Islamic' movements: those of Jamalu-ddin Afghani and the Caliph, Sultan Abdul Hamid Khan II. It was, too, the time of the 'modernist' movement in Islam associated with Afghani and his pupil Muhammad Abduh, Rector of Al-Azhar University in Cairo where many students from all over the Islamic world – including the Russian Empire, the Grand Duchy of Lithuania, and Poland – gathered and imbibed the new ideas. It seemed a good idea to stop the flow, and Russian influence in Persia increased as did its military pressure on the European ('Rumelistan') and Caucasian provinces of the Ottoman Empire.

It would seem that Russian imperial ambitions in what they still term the 'Near East' – that is, Turkey, Persia (Iran), and so on – were, until at least the October 1917 Revolution, aimed at the Anatolian and northern Levantine provinces of the Ottoman Empire together with Istanbul. Russian presence and influence in northern Persia seems to have been part of this, and also a means of trying to stem the two-way passage of men and ideas into and out of Central Asia and the Caucasus.

Thus, Russian policy toward the countries at the northern end of the Gulf has traditionally been defensive in character, aimed at shutting off the sources of ideas which have stimulated the Muslim subjects of the Empire. To do this effectively, it would have been necessary to overawe or physically occupy these states, or at the very least bring them within the Russian sphere of influence. Until 1980, this has only been possible in the case of Persia. It is probable that Russia's aim of simplifying internal stability within the Central Asian area by dominating the neighboring countries of the 'Near East' has been misread; leading to the 'warm-water port' idea. There has been circumstantial evidence to support this point of view, of course, such as the bid to construct a railway across Persia to the Gulf port of Bushire and the opening of a consulate there. But it is likely that the Gulf as such held little interest for them, until it became an area of strategic value to the West.

Muslim responses to Marxism

In the turbulent period before and immediately after the 1917 Revolutions, a number of other trends and events among the Muslim peoples of the Russian Empire contributed to the present attitudes of the Soviet Union toward Islam and the Muslims, which condition their perspective on the Arab and Muslim World. Among the leaders of what came to be called the *Jadid* Movement in Central Asia was Musa Jarullah Bigi (1875-1949). Bigi taught that the godless Bolsheviks were preferable to a militant Christian counter-revolution of the Whites and encouraged the Muslims to throw in their lot with Lenin and Trotsky. At the time of the February Revolution other leaders were encouraging their followers to support either the Mensheviks or Kerensky's Socialist Revolutionaries. During the Civil War many of them refused to take sides and remained neutral between the Whites and the Reds.

National consciousness was strong among all the Muslim peoples, and any set-back to Russian military prowess stimulated uprisings in the Muslim areas each acting as a trigger for the others. This consciousness was of a 'Pan-Islamic' nature. All the revolutionary Muslim reformers whatever their precise political views

had no doubt about the unity of the Muslim *ummah*, and they felt that revivification of Islam world-wide would come from Russian dominated Central Asia. Among these thinkers and activists, there had also developed an understanding of Marxism that was signally at variance with that of Lenin. The two most prominent Muslim communists were Mulla Nur Vahitov and the Tartar, Mir Said Sultan Galiev – 'the most brilliantly daring and influential National communist.'

Sultan Galiev propounded three essential ideas in his interpretation of Marx. Firstly, socialism could assume the character of the nation which adopted it. In the case of the Muslims, it should be compatible with and inside the framework of the teachings of Islam. Secondly, the world was divided into oppressors and oppressed. Non-proletarian nations could by-pass the capitalist stage of development and go directly from feudalism to socialism. Muslims, of whatever class of society, who had been the victims of colonialist oppression were therefore 'proletarians' and the national liberation movements in Muslim countries were socialist revolutions. The oppressors were the colonial powers, and the class struggle was between them and the colonial peoples. Thirdly, he rejected the Leninist dogma that socialism would solve the problem of different nationalities and replace them with socio-economic classes. Long experience of Russian domination caused the dismissal of the notion that Bolshevik Revolution would change the nature of Russian imperialism. He claimed that the proletariat of an imperialist nation which inherited power from the bourgeoisie was not basically any different from its predecessors and was bound to retain the old repressive colonial system although it might give lip-service to the rights of the colonized. Colonial exploitation of the Muslims would continue, and might even expand, under a future Russian-dominated communist regime.

Subsequent events have proved Sultan Galiev and the Muslim National communists to have been correct. Lenin predicted in 1913 'the right of every nation in Russia to political self-determination – the right, that is, of separating from Russia and of creating independent governments.' This was endorsed by the April 1917 Bolshevik conference in Petrograd (Leningrad). Attempts by Central Asia republics to establish their own communist governments, or to secede from the USSR, were suppressed by the Red Army. Russification has been pursued more vigorously under the USSR than it ever was under the Russian Empire. The intensity of Russian assertions to having a superior right to interpret Marxism for all seems

Above: A Soviet-supplied MiG-23 Flogger-E of the Libyan Air Force armed with AA-2 Atoll missiles, photographed in the mid-1970s.

Opposite page: Iranian Army supporters of Ayatollah Khomeini riding on a British-made CVRT Scorpion light tank through Teheran in 1979.

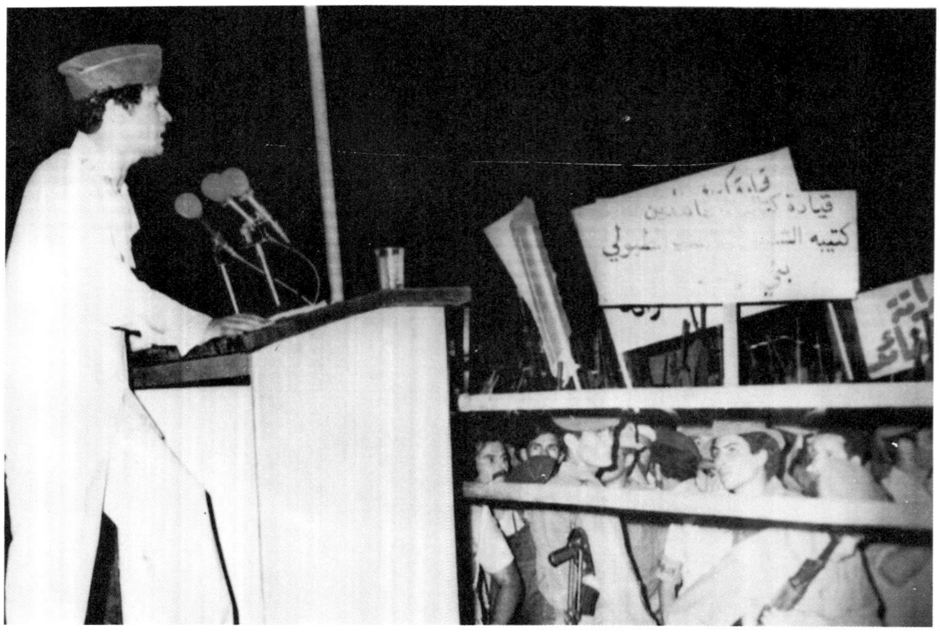

Above: Colonel Qaddafi, the Libyan leader, addressing soldiers near Tripoli in October 1977 to celebrate the anniversary of the Italian exodus.

Center page: Libyan women soldiers armed with Soviet AKMS assault rifles parade through Benghazi in 1979 to commemorate the tenth anniversary of the overthrow of King Idris al-Senussi in 1969.

to reflect an awareness of just how specious a claim this is. In considering socialist, and possible communist, regimes in the Arab and wider Islamic world, one should not be taken in by the Russian perspective. All the ideas and influences that led to the Jadid Movement and 'Sultan Galievism' are independently present in the Muslim world.

There was a time when the Muslims of the Russian Empire posed a direct threat to Russian leadership of the USSR. The two Pan-Muslim Congresses in Moscow (1 May 1917) and Kazan (July 1917) demonstrated a coherent Muslim National communist view of post-Revolutionary Central Asia, in which the Muslims of the Soviet Union constituted one nation. After 1923, they proposed the amalgamation of the entire Soviet Muslim world into the 'Republic of Turan' which was to be an independent and sovereign state of nearly 80 percent Muslims and speaking Turkic languages. To the Russian Bolsheviks this was totally unacceptable as the 'Republic of Turan' would have become the arbiter of world communism through its proximity to the colonial peoples and its Sultan Galievist-Marxist philosophy. They proceeded to break up the unity of the Soviet Muslims and purge their leaders, replacing them with Russians.

Communism and the Muslim World today

It is against such a background that one should view the Soviets' dealings with the Arab World and their attitudes toward Islam and the Muslims. Russian perceptions have developed over several centuries, and have consistently been reinforced. Ideologically, Soviet communism is seen both by the Muslims of the Soviet Union and the wider Islamic world and it would seem, by the Russians themselves as being a peculiarly Slavonic epistemology. There is also the ancient feeling of inferiority, admixed with the essentially political problems of ruling the Central Asian republics of the USSR. Nationalist sentiment has not abated there, and Russians display all the insensitivities and crude racialism of the more despicable colonial regimes. There also seems to be a Russian awareness not only of the burgeoning economic and demographic problems for the USSR that are arising in Central Asia, but also that the Soviet Muslims have not forgotten their past – whether that of the Islamic emirates and khanates, or that of Muslim National communism and Sultan Galiev – and that they are susceptible to influences and ideas that still reach them from the outside Islamic countries of the 'Near East' and the Arab World.

Thus, Soviet interest in the countries of Iran, Turkey, Afghanistan and Pakistan is in part similar to the Tsarist need to try and limit, if not wholly stop, the flow of ideas across the border. A watershed was reached in 1978 with the downfall of Shah Mohammed Reza Pahlevi in Iran and the success of the Islamic Revolution and the April communist revolution in Afghanistan which led to the Russian invasion in 1979.

Afghanistan

Initially, the Soviet forces sent into Afghanistan contained a large number of Central Asians who became 'intoxicated,' according to some commentators, with their exposure to a freer Islamic atmosphere. This repetition of the Czechoslovakia mistake of 1968 was speedily

remedied by their replacement with Russian and other European troops. However, in the meantime a brisk trade in clandestine purchases of Korans took place. There have even been reports of outbreaks of violence in Central Asia. In one case, it is said that a near riot developed when the Russian military authorities tried to stage an atheist burial for Muslim soldiers killed in Afghanistan. The local populace seized the bodies and gave them a proper Islamic burial.

It is thought that the Soviet troops in Afghanistan now number about 115,000 according to a 1984 estimate. This figure includes a number of MVD and KGB troops. The land forces' element of this large figure is thought to be made up of three Motor Rifle Divisions, one Airborne Division, one Air Assault Brigade, and an unspecified number of Naval Infantry cadres. Nearly all these persons are Slavs or other Europeans. Even Kazan or Volga Tartars, or Tartars from the former Grand Duchy of Lithuania, are weeded out of combat units drafted from the RSFSR or the Ukraine. Soviet Muslims serving in Afghanistan are restricted to support arms or construction units. Even then enough of them are being killed or injured in ambushes to create unrest in Central Asia. To Slavic parents burying their sons must come the brutal reality of the 'savage wars of peace.' Why should they pick up the 'white man's burden' in Afghanistan nearly a century after the British gave up? Afghanistan will not be the USSR's 'Vietnam', for the Indo-China War will not have changed American policy

Below: Soviet Motor Rifle troops in tropical uniform fraternize with Afghan Army commandos at a base during 1981.

Above: Afghan soldiers carrying a coffin draped in the Soviet flag that has just been unloaded from the Antonov An-26 Curl transport aircraft in the background some time in 1984.

Left: Soviet paratroops (blue berets) seen in Kabul with Afghan soldiers.

Opposite, top: The Russian-imposed Afghan leader Babrak Karmal talking to Soviet paratroops.

Opposite, bottom: Soviet Air Force Mi-24 Hind-A helicopter gunship returning to a makeshift airfield in Afghanistan. In the foreground can be seen a trailer-mounted Two Spot PAR (precision approach radar) and radio antennae connected to a Zil-157 radio truck. The six aircraft in the background are possibly Sukhoi Su-17/20 *fitters* for ground attack.

Above: Young Afghan boys seen toting an assortment of weapons (including a British-made Sterling sub-machine gun) in 1980. Such children go on to join the Mujahidin guerrillas when they are old and competent enough.

Opposite page: Afghan Mujahidin photographed at Barikot (Kunar province) just after Babrak Karmal came to power in December 1979. All are armed with Soviet weapons. There are four M-1944 carbines (two with folding bayonets extended), an AKM assault rifle, and a PPSh-41 sub-machine gun in evidence.

after the *trahison* of the peace movement has died away finally. However, Afghanistan may well stimulate something in the USSR which is truly revolutionary: the demise of Tsaristi-Bolshevik feudalism.

It is unlikely that the Afghan *mujahidin* can 'win.' It may be that in the long-term the mujahidin organisations can produce a single leader of the charisma of a Castro or a Sukarno, who can unite the country and its mutually hostile political groups behind him against the Russians. But even then, the mujahidin cannot regain their country if the Russians do not wish to give it back to them. The Russians can hold Afghanistan even if it means increasing the numbers of Soviet troops occupying the country. They can continue with this occupation as long as is necessary for them to achieve their aims; or indefinitely if they have lost sight of or changed their original purpose. The Soviet Union suffers from none of the restrictions of politics or ethics that limit the conduct of counter-guerrilla operations by Western nations. There will be no staying of the hand from the final blow as the Americans did twice after Linebacker I and Linebacker II in 1972.

Because of this freedom to do as they like, the Russians may stay in Afghanistan longer than they should for their own domestic well-being. Also, their presence in the country is having an effect on the parochialism of the Arab view of the world. This is changing from a concentric series of circles with Arabs (any sort) in the middle, then other Muslims, then the rest. The heightened awareness of Muslim *ummah*-consciousness that Russia's invasion of Afghanistan and Khomeini's Revolution in Iran have brought, are tending to revise this outlook into concentric circles of Muslim Arabs; other Muslims; Christians; and then the Rest. Neither Russia's nor America's relations with the Arab World will be the same after Afghanistan. In trying to replace the Hafizullah Amin regime with one which followed the Moscow line, instead of being content with one broadly sympathetic to Moscow, the USSR has started a train of events in Central Asia and its southern periphery that will probably have an unwelcome outcome for the Russians.

Central Asia

The Afghanistan invasion also brought the plight of the Central Asians back into the public eye of the Islamic world. The Iranian Revolution is more problematical for the Russians. Most Soviet Muslims are Sunnis, just as in the rest of the Islamic World, but the success of Ayatollah Khomeini's Shi'ite followers has acted as a widespread catalyst for the revival of self-confidence and an emerging consciousness of belonging to a bloc of nations separate from the communist and western worlds. This feeling is also spreading in Central Asia. Just as the defeat of the Tsarist armies in Manchuria in 1905 by the Japanese stimulated the founding of Muslim National communism, developments in

Above: A Soviet Army officer reads from *Red Star* to his vehicle crews while they take a break in Afghanistan. Their vehicles – a BMP-1 armoured personnel carrier and two MT-LB load carriers – are parked in a non-tactical line in the background.

Afghanistan and the Islamic Revolution in Iran may well produce a recrudescence of the Republic of Turan idea.

This is a considerable worry for the Russian communists, which could only be headed off – conceivably – by a spectacular victory in Afghanistan and a Moscow-orientated government in Teheran to undo the work of Ayatollah Khomeini. It was the perceptive American statesman Dean Acheson who said, just after the Second World War, that in the struggle against Soviet communism, the Muslims were the natural allies of the West. They are the most likely and best placed people to erode the power of Russian communism, even if it be through a different type of communism.

The growth rate of the Soviet Muslim population, and the residues of National communism and Sultan Galievism which they still hold to, have given the Russians yet another worry: *ozheltenie Krasnoy Armii* (the yellowing of the Red Army). The Soviet leaders are aware that the Armed Forces are the ultimate arbiters of the Soviet system, and if they are ideologically unreliable then the system will change. It is the perceived ability of the countries at the northern end of the Gulf to stimulate and reinforce these influences which probably causes significant anxiety in Moscow. The USSR's southern border holds a threat to the long-term survival of the present state of things greater than that given by any other front.

This old concern was probably appreciated by the wily Georgian dictator Stalin when he occupied Iranian Azerbayjan and Kurdistan in 1946 – a year before the *Zhdanovishchina* of the ideologue Andrei Zhdanov in September 1947 ushered in the Cold War, and before the area of the Gulf became such a crucial source of petroleum for Japan and the Western Europeans. Interestingly, the Soviet Muslims were then in favor of annexing the whole of Iran to the USSR, which would have considerably strengthened their position demographically and politically compared to the Russians. They probably have the same view today about Afghanistan. This is not what the Russians want. They would like a compliant buffer state whose government would discourage contacts across the border with the Soviet Union, and are not too fussy whether the regime would be Marxist or not.

The Arab World

It is a somewhat different matter with the progressive states of the Arab 'Rejectionist Front,' and with the various Arab communist parties. Here the concern is to control the 'line' taken so that it conforms with Moscow's perspective. This is more problematical than it may seem. Since the Soviet Union became the principal backer of Egypt in the late 1950s, and the PLO and certain other Arab countries in the 1960s, there have been many Arab Muslim visitors to the USSR. This has been accompanied by propaganda statements to assert or reinforce the Soviet Union's claim to be the 'historical and natural friend of the Arabs and Islam' – partly for domestic consumption by the Soviet Muslims. Contacts with these dignitaries have also had the effect of diffusing certain of their ideas. But Arab socialism and communism are of a different complexion from Russian Marxism-Leninism.

In the Muslim World, for an idea to take root it must be demonstrably in tune with the religion of Islam. If it can show historical predecents, especially those of the Prophet or his Companions, then it is likely to gain many supporters. It is for reasons of personal life-style and recorded statements which show a 'socialist' perspective that such Companions as Abu Dharr al-Ghifari or 'Umar ibnu-l Khattab have a significance far greater than that of Marx. So important is the Islamic dimension, that even the local Moscow-orientated communist parties have to adopt Islamic conventions. It has been recorded by eye-witnesses that at conventions of the Sudanese Communist Party, the meeting would begin with a recitation from the Koran, would break off at prayer times, and would finish with a supplication (*du'a*) to the Almighty. Speeches would contain frequent quotations from the Koran and the Traditions of the Prophet. Without all this, the communists would not have gained many followers, for the militant atheism of Soviet communism is well-known. The need to trim in this way, even if the leaders only mean it as a cynical gesture, has its effect on the development of communism in the Muslim world. It finds its own way to the Marxism of Mir Said Sultan Galiev. Thus, Arab socialism and communism are ideologically unsound for the Russians. Iranian versions are no better. Even the Moscow-line Tudeh party has been tainted by the popularity amongst the Iranians of the writings of Ali Shariati. But this is a belated realization by the Russians.

Because of the Russian and atheist connections of communism, the local communist parties in the Arab world have never made much headway. They have often been suppressed, and as their followers cannot convincingly explain away the 'scientific atheism' of Marxism-Leninism they have frequently been anathematized as godless heretics. It is difficult for a Westerner to comprehend the disgust and revulsion whch grips a Muslim when someone is branded as *Kāfir, bā-imān*, or *îmansiz* (all mean infidel). As a result, open communism is a liability and the Arab socialist parties have often seemed a better bet. But they too have a tendency to go their own way attempting to preserve a non-aligned posture.

Arab Socialism

Over all Arab socialism lies the long shadow of one man: Gamal Abdel Nasser. His achievements – at least as perceived and understood in the Arab world – and his political conduct condition to this day the programs and expectations of these parties. He is the inspiration for Mu'ammar al-Qaddafi of Libya and the paradigm for both Saddam Husain of Iraq and Hāfiz al-Assad of Syria, and was the mentor of Yāsir Arafat of the PLO. His brand of Arab socialism was fiercely nationalistic, non-aligned and anti-colonialist, but tried to remain within the modernist interpreptation of Islam.

Had the West – particularly the USA, Britain and France – acted with more sensitivity and understanding when the Young Officers seized power in Egypt in 1954 and afterward, then it is probable that Nasser's essential pro-Westernism would have produced a non-aligned but westward-leaning Arab world. It was, above all, the foolishness over the Aswan

Above: A convoy of civilian lorries waits for its military escort before facing the long and dangerous drive back to the USSR through the Salang Pass from Kabul in 1984.

High Dam project and the public humiliation of Egypt's envoy by Dulles that drove Egypt to the Russians. This was compounded by the flagrant breaches of the Tripartite Agreement by France in Israel's favor, Israel's rejection of Nasser's indications that he wanted a peace treaty, and the Israeli Gaza Raid (February 1955) which drove him to the Soviets for arms for Egypt's forces. Several contemporary sources indicate that Nasser did not want to rearm nor to fight a war with Israel at that time. But too much prestige was involved in the High Dam project to let American and World Bank refusal to fund it bring it to a halt. Also the public outcry at the destruction of the police post and the way the people lost their lives at Gaza could not be ignored. As the West would not help Egypt equitably, Nasser turned to the Russians. Not only did this allow the Soviets into the area to meddle in troubled waters and embarrass the NATO allies, but it also created a precedent and a system of supply developed as a result for Arab countries and movements to equip themselves for fighting Israel and the colonial regimes.

In all the colonial wars fought in the Middle East after the Young Officers' coup, Egypt featured as a backer and refuge. There was, and still is, a strong feeling of anti-colonialism in the Muslim world. This is not motivated by a sense of mission, as with Marxist-Leninist regimes,

but simply the wish to be rid of the foreign empires and their interference in the lives of the Muslims so that they can get on with their own civilization in their own way. This is partly the reason for the acclaim with which Nasser's attempts to play off the USSR and the West against each other were received. It also explains in some measure the tacit, but qualified, approval and support given to the regime of Ayatollah Khomeini in his outward rejection of both the West and the Soviet Union. One of the first steps that Nasser took was to buy the most powerful radio transmitter he could. *Sawtu-l'Arab* (Arab Voice) broadcast the success of the Egyptian revolution all over the Arab world and encouraged other Arab populations to emulate Egypt's example. A strong feeling at the time that the conflict between NATO and the Warsaw Pact was 'not our problem,' helped gain popular support for Nasser's campaign to break up the pro-Western Baghdad Pact. Constant calls to throw off the colonial yoke and also to remove rulers who were 'betraying the Arab cause' by siding too closely with the West had their effect, and one by one the Arab countries moved closer to Nasserism or had revolutions which brought in governments of that kind of persuasion. Even moderate monarchies have over the years tended to espouse Nasser's vision.

It is important to realize the extent and depth to which Nasser has affected the Arab world. Even people and parties which are avowedly anti-Nasserist, such as the *Ikhwānu-l Muslimoun* ('Muslim Brotherhood') in all Arab countries, or the Baath Parties of Syria and Iraq, have been subtly influenced by him. This is partly because in his appeals he was tapping generally-held and basic feelings and aspirations among the Arabs, and partly because during his career he raised the consciousness of the Muslims generally that they belonged to a civilization and culture that was older and richer than either the Western Christian or *parvenu* Soviet communist ones. Nasser's legacy has another effect. It is of the Arab socialism that owes nothing ideologically or politically to the USSR, but is their own. Those rulers, like Qaddafi, who appear to the West to be in Moscow's pocket should be reappraised.

Qaddafi simply takes Nasser's vision a number of stages further. The essential inspirations are Nasserism and an idiosyncratic understanding of the Koran. It is his own reading of the Koran, not *Das Kapital*, that has provided Qaddafi with his ideas. His exegesis is based on Nasserite Arab socialism taken to a possible extreme. It should also be noted that one of Qaddafi's inspirers, as with Nasser and Anwār al-Sādāt, is the original founder of the Muslim Brotherhood, Shaykh Hasan al-Banna. Some of the political views of the Muslim Brotherhood are 'Sultan Galievist.' It would appear that in Qaddafi's reading of world *realpolitik,* colonialism has not yet left the Arabs to get on with their civilization, but continues to interfere as before although in newer more subtle ways. This would seem to be Khomeini's reading too.

Above: A MiG-17 Fresco of the Syrian Air Force overflies advancing Syrian armored vehicles.

Opposite, top: President Saddam Hussein of Iraq greets the PLO leader Yasir Arafat when he arrived in Baghdad in November 1980.

Opposite, bottom: PLO fighters armed with Soviet-made AK-47 assault rifles photographed in October 1969.

If the situation is unchanged in essence, but different in form, then a similar alteration becomes necessary to the anti-colonialist struggle. Regimes which accommodate to the 'enemy' still need to be removed for betraying the Arabs, and armaments still need to be acquired. The only practical source of arms of the quality and in the quantities needed is the Soviet Union. But accepting arms from the USSR has never meant that the Arab regimes subscribe necessarily to Soviet communism, although the numbers of Russian advisers and instructors involved have usually produced KGB and GRU inspired subversion and instability in the country. The Russians need to bring the phenomenon of Arab socialism into line.

It was the extent of Russian interference in Egypt, and the way the USSR appeared to let the Arabs down when they needed support, that led to the expulsion of the Russians from Egypt by Sadat. Although many Arab countries have Treaties of Friendship and Cooperation with the USSR, it has not gone unnoticed that at certain crucial times the Soviet Union has not proved to be quite the friend of the Arabs that it sold itself as being. Although heavily dependent upon the USSR for military supplies, even the PLO, Syria, Iraq, and Libya, are wary about the reliability of their 'friend.' After the Egyptian experience, they have all moved away from the USSR. Libya buys arms from other countries – such as Brazil, Spain, Italy, France, Britain and even the USA – to diversify the supply and reduce dependence on the Soviet Union. Iraq, embroiled in a costly war with Iran and being funded by Saudi Arabia and certain Gulf countries, has made overtures to the West. Having been rebuffed, it has gone back to the USSR but is clearly unhappy with this. Syria, too, has been making similar moves. Only the People's Democratic Republic of Yemen (PDRY) seems content, but this is deceptive.

Above: Syrian soldiers photographed near Mount Hermon in 1973, armed with two RPG-7 anti-armor grenade launchers and a RPK light machine gun.

The Arabs and Israel

The key factor in the West's, particularly the USA's, inability to read and understand (or want to understand) the signs that most Arab countries wish to get their armaments from the West and not have much to do with the USSR after Egypt's experience, is the Israeli dimension. The continuing hostilities between Israel and the Arabs, the Camp David Agreement notwithstanding, have soured the Arabs' view of the West – especially America – and thrown the Arabs into a bad light in western eyes. What is seen as constant deference to Israel's wishes, whether or not these are in the real best interests of western countries, acts continually to drive the Arabs back to the Soviet Union. Exasperation at what they perceive as the injustice of this, leads even moderate Arab regimes to turn to the USSR and question the worth of 'friendship' with the USA. There is a sort of resignation abroad among the Arabs that whatever Israel does it will not bring forth even the slightest public rebuke from the USA. US Presidential elections are very public affairs, and the statements of the various candidates about Israel and the Arabs are taken to heart in the Arab world. It is this dimension more than any other which continues to permit the USSR to be an actor on the Middle Eastern stage after Afghanistan. Most of the Muslim world was appalled at the Russian invasion, and is anyway inherently suspicious about Soviet communism, so that it would welcome a way out of dependence on the USSR. Having largely given up hoping for a change from the USA, the Arabs are looking for more understanding and better things from the European Community.

The Egyptian dismissal was traumatic for the Soviet Union also. Its original involvement in the Arab world was to discomfit the European colonial powers and the USA causing a diversion of resources that might otherwise have been deployed along the Inner German Border, and also to try and break the encirclement of the USSR on its southern front. For a while it succeeded. Colonial wars exhausted the British and the French, and caused a political rejection of empire in the metropolitan center. The last rival to Soviet influence in the eastern Arab world and the Gulf – Britain – pulled out in 1972 leaving what seemed to be an open field. The Baghdad Pact had collapsed years before, and CENTO-RCD was no more than a residual notion. Arab support for the Eisenhower Doctrine was negligible, and Egypt seemed to be the pace-setter and arbiter of Arab politics. Constant American errors (as far as the Arabs were concerned) over Israel were driving them closer to the USSR at a time when the West needed to keep the Arabs on its side as it was becoming increasingly dependent on Gulf petroleum. The USSR must have misread its allies' dependence on it and the nature of Arab anti-colonialism, for the Arabs are not unaware of the Soviet Union's vast colonial empire of Central Asian Muslims. However, to them the USSR was not a power colonizing *Arabs*. It was the behavior of the Soviet Union in Egypt and elsewhere in the Arab world which made them see the Russians in the same light as they had previously viewed Britain, France and Italy. After being dismissed by Sadat, and the Syrians and Iraqis blowing 'hot and cold' toward the USSR, the Russians have come to consider the Arabs as being inherently unreliable.

Left: A Palestinian refugee camp in Jordan in 1969.

Below: Palestinians in south Lebanon in August 1980. They are manning two Yugoslav-made M-55 triple 20mm anti-aircraft guns mounted on Toyota Land Cruisers. All are armed with AK-47s.

The Yemen

Even the PDRY, which seemed to be the most convincingly Marxist-Leninist regime in the Arab world, has been undergoing changes. The problem of ruling turbulent tribes in a country with a strong feeling for the religion of Islam, had caused a moderation of the communism of the regime so that it is taking on a local color and having strong reminiscences of the National communism of the Soviet Muslims. The Islam is eroding the Leninism. Matters have deteriorated so far (from the Russians' point of view) that the PDRY not only called off its war with neighboring Oman, but also proposed a merger with the Yemen Arab Republic in 1981, and has been accepting Saudi Arabian blandishments. It is even arguable whether the Leninist nature of the regime's communism was ever more than a deception and whether it was not ideologically a militant version of Arab socialism.

The PDRY's armed forces are not very large. At an estimated total of 27,500 personnel (1984 figures), they are smaller than the Saudi Arabian National Guard. However, there is a significant number of military assistance personnel: 1500 from the USSR, 300 from Cuba, and 75 from East Germany. In addition, there are a further 500 Soviet personnel in the Yemen Arab Republic (YAR). It is thought that some of the PDRY's aircraft are flown by Soviet and Cuban crews. The small navy is mostly equipped with fast attack craft of various sorts: quite enough to close the Bab El Mandeb Straits from the Perim Island base. The larger air force – equipped with MiG-17Fs (highly maneuverable), MiG-21Fs and Sukhoi Su-20/-22s – is quite adequate to fly patrols over the naval blockade or over the ground troops. The army also possesses a surface-to-surface missile brigade armed with FROG and Scud B missiles.

A threat to the West?

Thus from a high point in the late 1960s, Soviet presence in the Arab world has declined significantly. Their influence is still very strong, and the Arabs need them for armaments. But this is really a factor of misguided US foreign policies, and the West's insistence on misunderstanding what the Arabs want. The USSR has succeeded in breaking the NATO encirclement of its southern front, by invading Afghanistan and by backing both Syria and Iraq, and also by helping Ayatollah Khomeini to power through the support given to him by the Tudeh Party. But what has replaced the previous situation in the Arab world is of a very much more mischievous nature for the Soviet Union in the long term. They have exchanged one Tsarist nightmare (encirclement) for another (infection of Central Asia by outside Islamic influence).

There are no established communist states in the Middle East and the Gulf – not at any rate in the sense of countries prepared to join with the Soviet Union in the 'international class struggle,' or which subscribe to the Marxist-Leninist view of the world. The only conceivable exception is the PDRY, the ideology of whose regime is suspect from the Russian point of view. Most Arab states, including the 'pro-Western' countries, will support Soviet interests to the extent that they differ with the USA over Israel. This is, however, very different from marching to Russian buglehorns on other matters. Countries like Libya and Syria may appear to be serving Soviet interests in their foreign policies, but these are policies that they have arrived at independently within their own outlooks and not taking any cue from Moscow. It is unlikely that they would support any Soviet adventures in a material sense. The only *quid pro quos* for past help from the USSR would be the use of ports and airfields for staging and R-and-R. In the event of a war between NATO and the Warsaw Pact, the Arab countries are likely to remain neutral and in a few significant cases side with NATO. None is likely to aid the Pact until they see how the war is faring, except to the extent of airfield and port facilities for emergency recoveries. The USSR is more likely

Opposite, top: Russian-made T-34 tanks being driven through San'a (Yemen Arab Republic) during a parade to celebrate the anniversary of the Yemeni Revolution.

Opposite, bottom and this page: Militiamen from south Yemen (PDRY) fighting to occupy the YAR border town of Al-Baida during a 1979 clash between the two neighbors. The picture below shows a complete AK-47 armed squad in an exposed firing position. The picture on the opposite page shows a detail of the PKM general purpose machine gun group located near the squad's left flank, a poor tactical position.

Above. Iranian Revolutionary Guards mounted on cross-country motor-cycles for mobile anti-armor patrolling. They are pictured in southern Iran on the Southern Front of the Gulf War in December 1982. They are armed with Soviet-supplied RPG-7 anti-armor grenade launchers and AKMS assault rifles.

to have free use of Ethiopia's Dahlak Islands than of Perim Island or the port of Aden (both PDRY).

No Arab or nearby Muslim country is likely to help the USSR in its imbroglio in Afghanistan. Apart from anything else, the domestic political cost would be too high. Of course, the Soviet Union itself could send forces or intervention forces to the Arab world to help. But it is unlikely that these would be asked for or welcome even in Libya or Syria. Military assistance is one thing, direct interference quite another, and it is probable that the Russians would find themselves involved in just the same problem as they now have in Afghanistan. Indeed, the Afghanistan invasion would be taken as an immediate parallel. The Arabs want support and supplies, not to exchange British (or French or Italian) colonialism for Russian. Their desire to be left alone culturally to get on with their own civilization applies equally to the Soviet Union as it does the West.

Soviet ambitions toward the Gulf are probably more complex now than they were a decade or so ago. Clearly, as an area of vital strategic value to the western European members of NATO, the ability to dominate it – if not deny it entirely to the West – would be a strategic aim of the Soviet Union. From this point-of-view, Russian airfields in the south of Central Asia and now in Afghanistan (such as Herat, Bagram, Kandahar) are important, as is the Soviets' ability to parachute or airland complete divisions of airborne troops and to deploy a great number of *vysotniki, reydoviki*, and *spetsnaz* teams. But Afghanistan has simply improved this capability, not initiated it. The Soviet Naval squadron in the Indian Ocean (normally about 25 ships strong) could certainly interdict shipping using the Gulf: but it would be a 'come as you are party' – it is unlikely that it could be significantly reinforced or resupplied in the event of a general war. Overland from the southern USSR, or Afghanistan, to the Gulf is across some extremely difficult and inhospitable terrain, and unlikely to bring the swift decisive victory desired by Soviet planners. Interference with this NATO asset is likely to be limited to

massive interdiction, not occupation.

As the world's greatest producer of petroleum, with large reserves both of petroleum and coal, it is not likely that the USSR will need other sources of supply for a few years. However, it has been suggested that by the end of the century, the Soviet Union will have to import some 20 million barrels a day. As the Gulf country with the largest reserves, Saudi Arabia, has only some 30 years' worth of petroleum left it is not likely to come from this source. For other reasons, if the USSR did have to import from the Gulf, it is probable that this would be done on some commercial or barter basis rather than by occupation or annexation. It is likely, too, that at least in the first instance the petroleum would come from Iraq's inland fields as was the case in the 1970s and natural gas from Iran (as was happening under Shah Mohammed Reza Pahlevi).

Because of the desire to insulate the Soviet Muslims of Central Asia from disturbing influences, it is unlikely that the Russians would want physical occupation of the Gulf. For to achieve this they would have to annex Iran, Iraq, the Gulf states and perhaps even Pakistan, and this would considerably exacerbate the very problem that they do not wish to make worse. The USSR is probably content with improving its ability to make the use of Gulf resources very expensive for NATO should it ever come to a war. It is also morbidly afraid of such a war, and is very worried that the Middle East generally and the Gulf in particular could provide the flashpoint for it. They are anxious that some reckless action by another outside power could precipitate the conflict and would like to talk with the West to defuse the tensions in the area and set up machinery for crisis management. The Soviets, no less than the Arabs and the Europeans, are worried about the behavior of Israel and what they see as America's reckless support for its adventures. There is good ground here for building up an understanding and mutual trust with the USSR and between NATO and the Arabs as well.

Mostly, it can be said, the Soviet Union has got itself into a situation that is inherently damaging. It went into the Middle East and the Gulf in the first place for sound reasons of *realpolitik*, but the internal contradictions of its own history and empire are catching up on it. Islam has proved through the centuries to be a very adaptable source of civilizations and political theories, and this has been shown to be so also with the socialism and communism of the nineteenth and twentieth centuries. Islam has colored and adapted these, too, into its own essential forms and within its own traditions. Arab socialism and Sultan Galievism spring directly out of this milieu, even if some of their ideologies rejected explicit adherence to the religion. Being compatible with Islam as a religion, these political theories find populations able to accept them as the basis of practical politics. Where, as in the USSR, Muslim peoples are at variance politically and philosophically with the colonial regime there is a dynamic which leads to resistance and rejection. This is happening in Soviet Central Asia, and has received stimuli from Russian involvement in the Arab countries of the Middle East and the Gulf, the Islamic Revolution in Iran, and the continuing problems of the Soviet invasion of Afghanistan.

The Russians just do not know what to do about the Central Asian problem. They cannot accept the programs which would defuse the situation. All they seem to do is to fall back on the Tsarist policy of trying to prevent the contact between the Muslims and the Iranians, Turks, Kurds and Arabs outside their borders being strengthened. After that, they just hope that Central Asia will go away. Their policy toward the Arab world has gone through a number of appraisals. This time, it would seem that what they strive for is control of Arab socialism to limit its impact and closure of the southern borders to stave off the Day of Reckoning with the Soviet Muslims. Imperial expansion that brings in yet more Muslims is not on the agenda.

6. AFRICA-EXPLOITING THE NATIONALISTS

Historical background

Prior to the 1950s, the African continent was primarily an arena from which all but European influence was firmly excluded. European control was itself a relatively recent phenomenon since, as late as 1870, not much more than 10 percent of Africa's land surface had been under direct European rule. By 1914, however, European influence had been extended over all but two states – Liberia, which had been established through American influence in the early nineteenth century; and Abyssinia, itself an aggressive imperialist power that had defeated an Italian attempt at subjugation. The so-called 'Scramble for Africa' between 1870 and 1900 had seen most of the great powers carving exclusive spheres of interest but not all. Imperial Russia for one had played no part in the division of Africa, preferring to seek expansion in the Balkans and the Far East. In China, of course, the decaying Manchu Empire was itself being partitioned by others, including both the United States and Japan.

The process of decolonization

In 1945 some 600 million people in Africa and Asia lived in territories that were not self-governing but the postwar years witnessed a rapid growth of a sometimes incoherent but nevertheless real demand by colonial peoples for independence. Such factors as the extension of western education and the concomitant development of indigenous elites; the development of an economic infrastructure, which through effecting 'de-tribalization,' westernization and urbanization helped to unify a colonial people in a physical sense; and racial tensions between white colonizers and the non-white colonial subjects did much to stimulate the nationalist upsurge. Above all, World War II acted as a catalyst not only in the Far East, where Japanese victories shattered white invincibility and prestige, but also in Africa. Many African servicemen encountered other nationalists and resistance groups when serving in the Far and Middle East. French Africa in particular was politicized by the divide between Vichy and Free French groups while some French colonies such as Morocco, Algeria and Tunisia were liberated by largely non-French armies. The Africans were conscious, too, that the Allied nations were fighting for 'freedom,' the United States' professedly anti-colonial line being as much a filip to nationalism as the growing awareness and understanding of the Soviet Union's ideological commitment to anti-imperialism. In a sense both emerging superpowers were thus tacitly in agreement on the undesirability of European colonialism. In any event, Italy, France, Belgium and the Netherlands had all been devastated by war while Britain was little more than an economic pensioner of the United States. But decolonization was a process which also involved as much a loss of will power as of real power, the predominance of liberal-democratic sentiments in most of the colonial powers (with the notable exceptions of Portugal and Spain) militating against the indefinite continuation of empire. Thus the colonial powers gradually acceded to the nationalists' demands although not always without a struggle.

Britain set the pace generally, divesting herself first of her Far Eastern possessions. There was some belief that African colonies were not yet ripe for self-determination but Ghana received its independence in 1957 and, in the wake of Prime Minister Harold Macmillan's 'wind of change,' Nigeria, Somalia, Sierra Leone, Uganda, Kenya, Tanzania, Malawi and Zambia had all followed by 1964. France had pursued different policies in her colonies, the Free French having announced at Brazzaville in 1944 that there would be a 'French Union,' in which all colonial peoples would enjoy full citizenship with direct representations in the French Assembly. However, nationalist uprisings against the return of colonial rule saw the French quit first Indo-China and then her North African colonies. The French offered a looser form of association but acceded to the preference of full independence, 14 new African states being created as a result in 1960 alone. By contrast again, Belgian rule in Africa had been paternalistic and devoid of any movement toward self-determination but, in 1960, the Belgians suddenly changed course and withdrew from the Belgian Congo in unseemly haste. Independence followed for Ruanda and Urundi two years later. Other African states such as Egypt and the Sudan, which had been protectorates rather than colonies, had also seen European withdrawal and by 1964 no less than 33 new African states had gained full independence.

Soviet and Chinese involvement

For the Soviet Union, spectacularly rapid decolonization offered new opportunities but also presented some problems. The Soviet's had the advantage, as did the communist Chinese, of being a 'clean' newcomer to African affairs untainted by the label of imperialists. But the emergence of independent Africa posed ideological difficulties for communist theory. Lenin had proclaimed the desirability of self-determination (provided this was not applied to non-

Russian minorities within the Soviet Union) but it was believed that nationalists would almost by definition be members of the bourgeoisie and not the proletariat. Thus, a second revolution would be required once the bourgeoisie had won independence. However, many African states clearly lacked any prospect of a developing urbanized proletariat and few of the new emerging African leaders professed to be Marxists. In fact, the Soviets had had little contact with any Africans prior to the 1950s except through the agency of others such as the French Communist Party, which had had links to nationalists such as Sékou Touré in Guinea. Indeed, the only communist party in Africa at that time was the white South African Communist Party, founded in 1921. A number of African leaders were also suspicious of the Soviet Union while virtually all post colonial economies were still closely tied to pre-colonial markets.

There was thus a strong element of pragmatism in the ideological recognition afforded by the Soviets to the concept of 'national democracies' in 1960 and to 'revolutionary democrats' in 1963, the hope being that those African leaders leaning toward socialism would effect a 'revolution from above.' Certainly the emergence of radicals such as Touré, Modibo Keita in Mali and Kwame Nkrumah in Ghana increased Soviet expectations but the initial Soviet approach was often tentative and appeared to be primarily aimed at securing diplomatic recognition for the German Democratic

Above: Former Patriotic Front guerrillas at one of the assembly points established in Rhodesia-Zimbabwe after the cease-fire of December 1979.

Aspects of the Soviet Union's involvement in the African continent: (*Right*) the guided missile cruiser, *Admiral Fokin*, and anti-submarine destroyer, *Vdokhnovenny II*, steaming into the Kenyan port of Mombasa in December 1968; (*Below*) A Soviet-supplied MiG-21 of the Ugandan Air Force flying past at a parade in January 1976 to celebrate Idi Amin's fifth anniversary in power; (*Far right*) More Soviet hardware supplied to Idi Amin, a 23mm anti-aircraft gun (*top*) and T-54 tanks (*below*).

739

Above: Female members of FRELIMO marching toward the Mozambique frontier in August 1973 from a base in Tanzania.

Republic and at achieving greater prestige. The problem was that not only did the idea that new states such as Tunisia and Algeria would be ideologically unpromising prove unrewarding, but the reliance placed upon individual leaders could equally result in diplomatic disaster. Thus, having failed to respond as quickly as the Chinese to the establishment of a provisional 'government' in Algeria in 1958 and having delayed recognition of Ghana, the Soviets chose to back the wrong horse in the (former Belgian) Congo. The support given to Patrice Lumumba was a turning point in terms of Soviet commitment to African affairs, the Soviets announcing in September 1960 that aircraft and pilots would be placed at the Congolese prime minister's disposal in his attempt to end the secession of the provinces of Kasai and Katanga. However, Soviet intervention aroused the suspicions both of the United Nations, which was heavily involved in the political morass of Congolese independence, and the Congolese Chief of Staff, Joseph Mobutu. Mobutu expelled the Russian embassy on 17 September 1960 when he moved against Lumumba. After Lumumba's death in February 1961, the Soviets backed the attempt by Antoine Gizenga to establish an alternative government to Mobutu at Stanleyville but this also foundered, Mobutu ultimately seizing full power in November 1965 following a protracted civil war involving numerous foreign mercenaries, in what became Zaire.

Other failures followed from too close an identification with radicals, Nkrumah being ousted in a coup in 1966 and Keita in 1968, while an attempt to meddle in the internal affairs of Guinea led to Touré refusing to permit Soviet aircraft to refuel there during the Cuban Missile Crisis in October 1962. In Kenya the Soviets similarly suffered when their protogé, Oginga Odinga, left the government for opposition in 1966 and was subsequently arrested in 1969. Nor was Soviet economic assistance to Africa always successful in reaping rewards since the Soviets were less generous than the Chinese in terms of interest chargeable on long-term credits and Soviet machinery, often provided in return for primary products such as Ghanaian cocoa, was not always reliable. There was frequent mismanagement of aid projects and considerable waste of resources while Soviet personnel acquired a reputation for aloofness in contacts with Africans. Soviet aid was extended to some sixteen African states by 1966 and a wide variety of educational opportunities also made available through such agencies as the significantly named Lumumba People's Friendship University in Moscow. However, Soviet aid was miserly in comparison with that extended to other areas – between 1957 and 1974, for example, only some $780 million worth of aid went to the countries of sub-Saharan Africa while $850 million was the total sum expended on aid to Afghanistan alone in the same period.

More concrete successes derived from the supply of military equipment; indeed, the very first Soviet contact with the continent was the supply of arms to Egypt in 1955 through the agency of Czechoslovakia. Ironically, Soviet success often resulted from the failure of the West to respond when initially approached for weapons. This was the case in Guinea in 1960, in Somalia in 1963 and in Nigeria in 1967-8. In Somalia, which had become independent in 1961, none of the western powers was prepared to boost Somali territorial ambitions which threatened American interests in Ethiopia, French interests in Djibouti and British investment in Kenya. The result was that the Soviets easily outbid the West, aid being extended in November 1963 and the first advisers sent in 1967. Soviet influence was further increased by a left-wing coup in Somalia in 1969. An even greater success was that in Nigeria since some western interests were involved in the secession of Biafra from the Nigerian federation in May 1967. The United States preferred to remain strictly neutral and Britain refused to give the Nigerians the arms they demanded to crush the Ibos of Biafra.

Above: Zairean paratroopers of the Inter-African Peace Keeping Force arriving at Ndjamena in Chad in November 1981 while, in the background, Libyan troops line up to board their Soviet Ilyushin planes.

Having failed to secure jets and tanks in London, the Nigerians went on to Moscow. Assured of American disinterest in the conflict, the Soviets supplied first through the Czechs and, then directly, 12 Czech Aero L-29 Delfin jet trainers, 10 MiG-17s, six Mig-15s and, in April 1968, three Ilyushin Il-28 bombers. Some 122mm artillery was also supplied while the aircraft were mainly piloted by Soviet-trained Egyptians.

In fact, the Soviets had been favorably disposed toward the Ibos, whom they regarded as more progressive than other tribal groups in Nigeria. They were also wary of antagonizing states such as Tanzania and Zambia that supported Biafra. On the other hand, the Nigerian government was recognized by the Organization of African Unity (OAU), founded in 1963, and the Soviets were always careful to acknowledge the OAU's aims. But even success in states such as Somalia and Nigeria could prove fragile and ultimately the Soviets were to be expelled from both in 1977 and 1979 respectively. That fragility was a reflection of the way in which Soviet gains in the 1950s and 1960s largely depended upon the willingness of Africans to accept the Soviet presence. On balance, the Soviets had little real success in the first two decades of their involvement in Africa, the apparent nadir of influence being reached in July 1972 when the Soviets were expelled from Egypt, the only country on the African continent in which Soviet combat personnel have been actively deployed (as missile operators and pilots from 1967 to 1972).

By contrast to the Soviet Union, the People's Republic of China enjoyed the advantage of being rather farther away from Africa and a source of apparently disinterested but effective support for newly independent states. The Chinese also believed, with some justification, that their concept of the need for continuing anti-colonial revolution was far more attractive to the Africans, who often deeply resented their former colonial masters, than the Soviet version of communist orthodoxy. The Chinese likened African development to a previous period of China's own history and thus also contrived to appear as having a ready model for Africans, which might be more appropriate than that offered by the Soviets. There were, of course, benefits to be gained for the Chinese in boosting the prestige of Mao Tse-tung in playing a role on the global stage as well as the possibility of cultivating economic links. The Chinese were altogether more prepared to invest in long-term expectations of social revolution than the Soviets and were far more flexible in their general approach. Chinese aid was also more generous with loans on longer terms, often interest-free or even outright gifts. China was also better placed to play the 'Afro-Asian' card since it was not only non-white but underdeveloped and could pose more effectively as non-aligned and neutralist than the Soviets, who were clearly involved in ideological confrontation with the United States. Thus the Chinese became closely linked with the Afro-Asian Conference at Bandung in Indonesia in 1955, at which six African states were represented, and participated in its off-

Above: The Chinese premier, Chou En-lai, accompanied by President Julius Nyerere, inspecting a guard of honor at Dar Es Salaam on his arrival in Tanzania in June 1965.

shoot, the Afro-Asian People's Solidarity Organization (AAPSO) and its various agencies such as the International Committee for Aid to Algeria and the Congo and the Afro-Asian Solidarity Fund Committee.

China therefore enjoyed some early success, beginning with economic links operated with Egypt in 1955 whereby the Chinese undertook to buy cotton that western states had refused to buy at a price acceptable to the Egyptians. China was also quick to recognize the provisional 'government' in Algeria in September 1958 and by 1970 had diplomatic relations with fifteen African states at a time in which many western states still regarded the Nationalist Chinese on Taiwan as the legitimate government of the whole of China. But there were also setbacks, five other African states having broken off relations with Peking by 1969. Rather as Soviet influence had waned when individual leaders were ousted, so China suffered similar events such as the coup in Burundi in 1965 and in the Central African Republic in 1966. Nkrumah was actually visiting Peking when he was toppled in February 1966 and the Chinese were left with the thorny problem of balancing relations with Nigeria's new rulers against the support for Nkrumah that persisted in other states such as Mali and Guinea. A similar problem had occurred in the previous year when, in order to save the planned holding of the second Afro-Asian conference in Algiers, the Chinese had precipitately recognized the government of Colonel Boumedienne, who had ousted their former friend, Ahmed Ben Bella, in a coup in June 1965. Coups generally were anathema to leaders of newly independent African states and Chinese haste undoubtedly damaged her standing. The Chinese had invested considerable effort in working toward the conference, their foreign minister, Chou En-lai having made extensive tours of African states in 1963 and 1964. The collapse of the conference was a blow, the Chinese subsequently losing interest in AAPSO, and it was therefore understandable that there was hesitation after Nkrumah's fall. The doubts resulted in the new Nigerian government breaking off relations with Peking in November 1966 and the Chinese subsequently backed Biafran secession. Other setbacks also occurred in Tunisia and Kenya while the Chinese were also involved in backing opposition groups in Malawi, Niger, Rwanda and Zaire.

The Cultural Revolution from 1966 to 1969, during which Peking had only one overseas ambassador in post, also damaged Chinese standing in Africa but a major success was the undertaking of the Tan-Zam railway linking Zambia's copper belt with the port of Dar es Salaam in Tanzania. The Chinese opportunity derived from the failure of the West to offer similar assistance in freeing Zambia of reliance upon railways through white Rhodesia and the Portuguese colonies of Angola and Mozambique. Preliminary survey work was undertaken from August 1965 and a report submitted in October 1966, an agreement being signed in September 1967 for the surveying and designing of the line at an estimated cost of $14 million.

Above: President Kwame Nkrumah of Ghana flanked by the Chinese President and Premier on his arrival in Peking on 24 February 1966. Nkrumah was deposed by a coup during his visit.

Construction began in 1970 with a Chinese loan to be worth $336 million and the railway was completed in mid 1975. China undoubtedly weighed the political advantages of undertaking the project when others declined to do so and in being able to complete the engineering required. An economic project of this nature also demonstrated China's lack of interest in directly intervening in the internal political affairs of African states, while actually subtly advancing Chinese interests and the prospect of Zambia and Tanzania participating more actively in the struggle for southern Africa. China was also consciously rivalling the Soviet Aswan Dam project while her technicians, though isolated from the local population, were extremely conscientious in their dealings with African workmen and were seen to be making demonstrable sacrifices on their behalf.

While the Chinese were cultivating generally more successful links with Africa in the 1960s, the Soviet attitude toward the third world as a whole underwent something of a change, most notably after the fall of so many 'revolutionary democrats.' The Soviets were now more conscious of the lengthy process which might be involved in the transition to 'genuine socialism' and an even higher degree of pragmatism entered their relationships with Third World states. Thus, by 1971, diplomatic relations had been established with 32 out of 41 independent African states and the Soviets went on to establish links with even markedly pro-western states such as Botswana and Djibouti in the late 1970s. They were also concerned to place economic links on a sounder financial footing. Trade with Africa since 1970 has not been of great importance to the Soviet Union but there have been some mutually beneficial projects such as the exploration for and exploitation of raw materials such as bauxite from Guinea and Algeria, the latter being one of the few minerals with which the Soviet Union itself is not already plentifully endowed. Links have also taken the form of fishing agreements with 23 of the 35 maritime states of the continent (by 1980) while the Soviet airline, Aeroflot, has steadily increased its African network.

To a large extent, however, such links have also been pursued on purely strategic grounds, Soviet strategic interests being given priority irrespective of the nature of the leadership or system in an individual African state. Increasingly, from the 1960s onward, the Soviet Union was developing a naval capacity which it had previously lacked. Partly, the new interest in naval affairs was the result of the extension of the United States' ability to threaten the Soviet homeland from, firstly, carrier-borne forces and, secondly, submarines operating in the Mediterranean and the Indian Ocean. Polaris submarines entered the former in 1962 and the latter in 1964 while the Cuban Missile Crisis of October 1962 demonstrated the Soviet lack of capability to deal with naval blockades. Partly, too, the growth of the new Soviet Navy was inspired by the traditional correlation of seapower with global status and as a highly visible means of projecting power and capability. Thus Soviet naval vessels first appeared in the Indian

Above: Soviet-equipped Tanzanian troops on parade in April 1979 at a time when the Tanzanian army was employed in the overthrow of Idi Amin.

Top right: An East German photograph of a session in an African National Congress guerrilla training camp. East Germans have advised many African guerrilla groups including FRELIMO and ZANU.

Bottom right: An SA-9 S(urface) to A(ir) M(issile) system captured by South African forces during operations against SWAPO in Angola.

Ocean in 1968 and between 1968 and 1971 made a total of 54 visits to East African ports (32 of them to Somalia) and 42 visits to Red Sea ports. In 1969 Soviet vessels appeared off the Atlantic coast of Africa and in 1970 patrolled off the coast of Guinea in the aftermath of an attempted invasion by Portuguese-backed exiles.

Once the Soviet Union had begun to develop its naval and air transport capabilities, advantage accrued in terms of the possibility of intervention in more distant conflicts. In earlier years the Soviets had been anxious to avoid involvement in local wars which might provoke confrontation with the United States and lead to more generalized war, even nuclear war. By the early 1970s, however, there was new optimisms in Soviet military circles that such involvement might be achieved not only at low cost but relatively free of risk in an era of détente. Some benefit would also be derived from being seen to be able and willing to act in certain situations, especially if this weakened Chinese influence, but it seems unlikely that rivalry with China was the sole or even an important determinant in Soviet willingness to seize what opportunities might be presented. In a sense, too, the greater success the Soviet's appeared to achieve in the 1970s in Africa was because those opportunities were so presented either through western weakness or through western failure to realize African expectations. Two critical areas were the question of nationalist movements in southern Africa and the political developments in the Horn of Africa.

Direct assistance for the Nationalists

Soviet support for 'wars of national liberation' has been manifest ever since Nikita Khrushchev's announcement to that effect in January 1961. In Africa in particular, this was a useful card to play given the West's identification, first, with colonialism and, secondly, with apparent support for the racial policies of the government of the Republic of South Africa. From the very beginning, therefore, the Soviets were involved in extending assistance of one kind or another to nationalist guerrilla groups in southern Africa. In the Portuguese colonies of Angola, Guinea-Bissau and Mozambique, nationalist guerrilla action commenced in 1961, 1963 and 1964 respectively. In Guinea-Bissau the Soviets supported the group known as *Partido Africano de Independência da Guiné e Cabo Verde* (PAIGC); in Angola, *Movimento Popular de Libertacão de Angola* (MPLA); and in Mozambique, *Frente de Libertação de Moçambique* (FRELIMO). However, in each colony there were rival nationalist groups, the divisions being largely along tribal lines, and the Chinese were also given the opportunity of sponsoring nationalism. In Angola, China backed the *União Nacional para a Independência Total de Angola* (UNITA), which had broken away in 1966 from yet another group known as *Frente Nacional de Libertação de Angola* (FNLA). The Chinese also supported the breakaway group of *Comité Revolucionario de Mocambique* (COREMO) in Mozambique but the divisions were not always

clear cut. Thus, both PAIGC and FRELIMO received largely Soviet weaponry but adopted Chinese theories of rural guerrilla warfare which were more applicable to the nature of the conflict. Although the Chinese lost influence in PAIGC by 1969, Chinese military theory still found favor with PAIGC leaders. Similarly, Chinese instructors continued to assist the training of FRELIMO in the refuge of Tanzania although the FRELIMO leader after 1969, Samora Machel, had been trained in the Soviet Union and Algeria. Some aspects of FRELIMO's internal organization also continued to reflect Maoist ideals.

In many cases the guerrillas utilized a wide range of weaponry from a number of communist sources. PAIGC, for example, used the ubiquitous Soviet AK-47 assault rifle as well as other Soviet weapons such as the Shpagin machine gun, RPG-2, M1943 7.62mm machine gun and M-14 mine. By March 1974 Soviet supplied surface-to-air missiles (SAMs) were also beginning to appear in Guinea-Bissau and PAIGC had use of Soviet 122mm rocket launchers donated by Nigeria. But PAIGC also used Cuban LGF 89mm bazookas and Chinese 75mm Type 52 recoilless rifles, 120mm mortars and claymore mines. Actual training of guerrillas also took place in many communist states including the Soviet Union and China, much training in the former being undertaken at locations such as the Komsomol School in Moscow and the guerrilla warfare school at Simferopol in the Crimea.

Right: Soviet AGS-17 30mm grenade launchers captured by the South Africans inside Angola.

Far right, top: Louis Cabral, President of Guinea-Bissau and Deputy-President of PAIGC in conversation with Party members in June 1974, shortly after independence.

Far right, middle: A PAIGC commander issuing instructions to guerrillas during the campaign against the Portuguese in Guinea-Bissau in August 1971.

Far right, bottom: Largely Soviet-equipped PAIGC guerrillas photographed by a Bulgarian cameraman in March 1974.

The rivalry that existed between Soviet and Chinese promotion of different nationlist groups was equally apparent elsewhere in southern Africa. In the case of Rhodesia, which declared UDI from Britain in November 1965, the Soviets backed the Zimbabwe African People's Union (ZAPU), led by the veteran nationalist Joshua Nkomo. By contrast the Chinese favored the Zimbabwe African National Union (ZANU), which broke away from ZAPU in 1963 and was led after 1974 by Robert Mugabe. Again, there was apparent evidence that Chinese rural warfare theories better suited African conditions, Nkomo refusing to commit the bulk of his forces to the guerrilla struggle inside Rhodesia because he favored holding them back for a Soviet style conventional assault. This may well have cost Nkomo the advantage of possession of ground when the war in Rhodesia ended in 1979. Once more, others were also involved in sponsoring revolutionary guerrilla activity. Of those nationalist guerrillas killed inside Rhodesia between 1966 and 1968, a total of 52 had been trained in Tanzania, 35 in Cuba, 34 in the Soviet Union, 22 in Algeria, 14 in China, 5 in Czechoslovakia, and 4 in Eygpt. East Germans were also active in fighting beside ZANU guerrillas during the successful Rhodesian raid ('Operation Miracle') on New Chimoio in Mozambique in September 1979.

In the territory of South West Africa, known to the nationalists as Namibia, the Soviets backed the South West Africa People's Organization (SWAPO) while the Chinese supported the South West African National Union (SWANU). As in the case of UNITA in Angola, known as the 'Black Chinese,' SWANU was as concerned with political action as military action and did not, in fact, take up arms against the South African administration of the territory as SWAPO did in 1965. The division of communist sponsorship also extends to South Africa itself where the Soviets support the African National Congress (ANC) and the Chinese the Pan-Africanist Congress (PAC), both groups being banned by the South African government since 1963.

As already indicated, Soviet or Chinese investment in nationalist guerrilla wars was low in cost and in risk while holding out the prospect of gain in the event of any of their protégés coming to power. In the event, the whole situation was utterly transformed by the sudden and totally unexpected collapse of Portuguese resolve in maintaining their African colonies. Portugal was far from military defeat in Africa in 1974 although the security situation in Mozambique was deteriorating. In Angola the war had become one of low level stalemate while the most successful group, PAIGC in Guinea-Bissau, had also been largely contained after 1968. The Portuguese retained control of all major routes and all urban centers and, in a curious way, it could be argued that the colonial campaigns had actually sustained the regimes of Antonio de Salazar and his successor, Marcello Caetano, through military expenditure stimulating the backward Portuguese economy in the 1960s and having promoted a semblance of national unity. In short, no one had predicted any real slackening of Portugal's effort to stay in Africa. However, a combination of domestic social, political and economic factors in Portugal itself and mundane professional grievances within the Portuguese armed forces, when coupled with the inevitable strain of 13 years of continuous war, led to a military coup by the so-called Armed Forces Movement (MFA) against Caetano on 25 April 1974. The first indication was that the views of the former Commander-in-Chief in Guinea-Bissau, General Antonio de Spinola, would prevail and that Portugal would seek some kind of federal solution to its future relationship with the colonies. Within a short time left-wing elements within the MFA ousted Spinola and opted for withdrawal from Africa.

In Guinea-Bissau and Mozambique, it was relatively easy to hand over power quickly to PAIGC and FRELIMO respectively since no other nationalist groups had lasted the course. Negotiations in Guinea-Bissau in May 1974 led to complete Portuguese withdrawal by October while, in Mozambique, agreement was reached in September 1974 on a program for independence by June 1975. In Angola no such solution presented itself for the three rival nationalist groups of MPLA, FNLA and UNITA were all still very much in the field. The Portuguese patched together a makeshift coalition government to share power in January 1975

but fighting had broken out long before the projected date set for independence on 11 November 1975. Matters were further complicated by the intervention of external agencies in order to secure success for one group at the expense of the others. The United States had been covertly channelling funds to Holden Roberto's FNLA ever since 1962 as a kind of insurance against Portuguese withdrawl since Roberto was regarded as a moderate. In January 1975 the United States Central Intelligence Agency (CIA) resolved to advance $300,000 worth of assistance to FNLA. China was also involved in Angola through support for UNITA. Following a reconciliation with Mobutu, whom the Chinese had once described as a 'neo-colonialist swindler and puppet,' they had also sent 112 advisers to Zaire in May 1974 to help train FNLA forces. Zaire had long been an active supporter of FNLA, Mobutu being Roberto's brother-in-law, and Mobutu also coveted the oil fields of the Cabinda enclave. The number of Chinese advisers increased to about 250 by July 1975 although the Chinese then began to withdraw for fear of association with South Africa, which also supported the alliance forged between FNLA and UNITA. The Chinese were to claim subsequently that they had initially attempted to encourage agreement between the three rival groups in Angola. A limited number of South African troops had been active in some southern areas of Angola in cooperation with the Portuguese for some years, guarding vital installations such

as the Cunene hydro-electric project and it would appear that the South Africans were led to believe that, at the very least, they would receive tacit United States support for any actions they took to assist FNLA and UNITA in the developing civil war. Direct South African assistance to these groups began in August 1975, possibly as early as July, although large scale intervention did not occur until October 1975.

The MPLA was also in receipt of external assistance, the pivotal role being played not by the Soviet Union but by Cuba. Cuba had extensive recent history of being involved in Portuguese Africa, a Cuban being captured by the Portuguese inside Guinea-Bissau in November 1969 and four more in the course of the following year. Cuban troops had appeared in Africa as early as 1962 when two battalions had assisted Algeria in its brief border war with Morocco and, by 1970, Cuban advisers had also been visible in Guinea, the former French Congo or Congo-Brazzaville, and the Cameroons. The Cuban image was of an underdeveloped Third World country, prominently involved in the 'non-aligned' movement. The Cuban leader, Fidel Castro, had also turned his attentions to Africa through the failure to carry the Cuban revolution to the remainder of Latin America in the 1960s, Africa being represented as a new challenge. Undeniably, the Cubans also saw themselves as 'Africans' since some two-thirds of the population were of negro or mulatto descent, many allegedly actually coming originally from Angola as slaves. Castro himself spoke in December 1975 of Cuba being a 'Latin-African nation' and, significantly, the Cuban intervention was to be code-named 'Operation Carlotta', after a female black slave who had died while leading a revolt against Spanish rule in the 1840s.

There were Cuban advisers with MPLA forces long before 1975 but the Cubans subsequently claimed that they were approached by the leader of MPLA, Agostinho Neto, for more substantial support in May 1975. According to the Cuban version of events, a mission was dispatched to Angola in July and 700 to 800 troops arrived in the first eleven days of October 1975. The Cubans also claimed that no decision was taken to commit larger numbers to assist MPLA until early November but it seems certain that the crucial decision to do so was taken in early October. It does appear, as the Cubans later said, that they would have acted anyway whatever the attitude of the Soviet Union with regard to intervention. The Soviets had, in fact, cut off aid to MPLA in 1973 when an internal split in the nationalist ranks developed between Neto and Daniel Chipenda. The Soviets only resumed aid in late 1974, probably because of the support that China was now giving to FNLA, but there is little doubt that material assistance developed rapidly in the early part of 1975. In April a Yugoslav freighter docked at Luanda in Angola and offloaded trucks and SAM-7 missiles before being sent away by the still resident Portuguese authorities to Pointe Noire in Congo-Brazzaville, where the remainder of the cargo was unloaded. In the next few months a series of Eastern Bloc vessels docked either at Luanda or Pointe Noire, supplies landed at the latter being re-shipped to Angola in Soviet-supplied landing craft. In March and again in August 1975 military equipment was airlifted by the Soviets into Congo-Brazzaville and then flown to airstrips held by the MLPA inside Angola, a process which led the United States to step up its own financial and material assistance to FNLA and UNITA to a vaue of $10 million. In turn, the Soviets then began to supply heavier equipment including tanks and artillery and, in all, it is believed that at least $200 million worth of Soviet equipment reached MPLA between March and December 1975.

The situation in Angola during 1975 was thoroughly confused and there are conflicting versions of events and of the responsibility for those events. An important development was undoubtedly the advance into northern Angola of a joint FNLA and Zairean column in March 1975, which appears to have prompted the first major Soviet airlift. The success that MPLA then enjoyed, in turn, engendered a response from the United States both in terms of increased covert and overt assistance and from South Africa. In October 1975 the South Africans intervened more decisively with four columns, codenamed Zulu, Foxbat, X-Ray and Orange, which, in cooperation with UNITA forces, drove toward Luanda from the south while the FNLA and Zaireans advanced from the north. It was at this point that Cuban troops began to arrive in large numbers by air and by sea with equipment that included 80 T-34 and 10 T-54 tanks and over 100 of the devastating 122mm rocket launchers that played an important part in defeating the FNLA and Zairean forces in the north. Some 400 Soviet advisers also arrived during November while Cuban strength rose to some 2500 men. A total of some 11,400 Cubans had arrived by January 1976 and Castro later admitted that the level of Cuban strength in Angola was to peak at 36,000 troops.

Despite the Cuban arrival, there were few direct clashes with South African forces, the latter's advance being spearheaded by armored cars, 140mm howitzers and 81mm mortars. The most serious confrontation between the two forces was at Bridge 14 from 9-12 December 1975, in which the South African 'Foxbat' column inflicted losses including an estimated

Above: UNITA forces in training at Nhuha near Silva Porto in Angola in October 1975. UNITA continues to challenge Angola's Marxist regime.

Far left, top: Soviet T-34 tanks manned by Cuban troops occupy a main road in the Angolan town of Huambo after Cuban intervention in the Angolan civil war in 1975.

Far left, bottom: A Cuban soldier driving a tractor on a nationalized farm in Angola. The Cuban presence remains substantial.

Above: Angolan and Soviet naval officers confer together. The success of Soviet initiatives in Africa has opened ports in the Atlantic and Indian Oceans for their Navy.

Top right: MPLA supporters in Huambo in Angola in 1976 celebrating the victory secured by Cuban intervention.

Bottom right: Military supplies being unloaded from a Soviet vessel at Luanda in Angola during 1984, an indication of the MPLA's reliance on the Soviet bloc.

200 dead on Cuban forces for the loss of only four South African lives. Less than 2000 South Africans were actually committed to Angola but they still got within 200 miles of Luanda before the increasing build up of Cuban backed opposition brought their advance to a halt. The South Africans later claimed that they had been 'left in the lurch' by the United States. In fact, the US administration or rather the Secretary of State, Henry Kissinger, had been less than honest in dealings with Congress over the extent of United States assistance to FNLA and UNITA. In the post-Vietnam mood of opposition to foreign involvement that might ultimately cost American lives, and in the post-Watergate mood of Congress, endeavoring to wrest back the initiative in foreign policy from President Ford's lame duck administrative that had already lost the November election, Senate voted on 19 December 1975 to cut off all military and financial aid to FNLA and UNITA. The House of Representatives endorsed the decision in February 1976 and the new Carter administration even indicated in June 1976 that it would end assistance to Zaire as well.

Thus the airlift and sealift of troops and matériel effected by the Cubans and the Soviets ensured victory for MPLA, the South Africans withdrawing by March 1976. The airlift itself had been accomplished both with ageing Cuban Bristol Britannia aircraft and modern Soviet Antonov An-12 and An-22 transporters refueled in the Azores. Other countries had refused refueling facilities including Trinidad and Barbados in the West Indies, although up to 5000 Cubans may actually have passed through Barbados in civilian clothes aboard aircraft refueled there before the Barbadian government became aware of what was happening. Later flights were therefore routed through the Cape Verde Islands as well as the Azores. However reluctantly the Soviets may have become involved in the intervention, it was undeniable that many advantages accrued from it by contrast to western indecision and weakness. Success for the MPLA projected Soviet power and influence through the agency of the Cubans, who were regarded as 'surrogates,' into the heart of southern Africa. Both Mobutu in Zaire and President Kenneth Kuanda of Zambia were particularly alarmed by what the latter referred to as 'a plundering tiger and its deadly cubs coming in through the back door.' The establishment of Soviet influence in Angola also served to open up the possibility of developing Angolan ports as maritime bases in the South Atlantic and of making available the rich mineral and agricultural resources of Angola. More importantly, the Soviets had shown themselves willing and capable of acting in support of their friends while the Chinese and the West had not done so effectively. In the

MPLA
LEGÍTIMO
REPRESENTANTE
DO
POVO
ANGOLANO
INDEPENDÊNCIA COMPLETA
UM SÓ POVO
UMA SÓ NAÇÃO
4FEV 61 76
15 ANOS DE LUTA ANTI-IMP

Inset, below: Soldiers of the Western Somali Liberation Front during the Ogaden War of 1977/8, the conflict that forced the Soviets to choose between two Marxist states.

Main picture: Soviet tanks of the Somali Army being paraded in Mogadishu prior to the break in relations with the Soviet Union's first real ally in sub-Saharan Africa.

process, the West had apparently allied itself openly with South Africa, an arrangement still anathema to most black African states. The Soviet gamble in risking United States retaliation had been more than justified by the failure of resolve in Washington and, to the Soviets, Angolan success more than compensated for recent failures in the Middle East.

Further opportunities for the advancement of Soviet interests derived just two years later from the political situation that developed in the Horn of Africa. As already mentioned, Somalia had represented a long-term investment for the Soviet Union in an area of great strategic significance close to the Middle East and the Persian Gulf, from whence the West derived much of its oil. Some $450 million worth of Soviet economic and military aid was poured into Somalia by 1977 and the Soviets, who had greatly escalated arms supplies after 1969, also made the unparalleled gesture of cancelling a Somali debt of some $45 million. A treaty of friendship had been concluded with President Siad Barre of Somalia in 1974 and the Soviets were developing naval facilities at Berbera, communications facilities at Mismayu and airfields at Mogadishu, Hargeisa and elsewhere. But Somalia had expansive aims of its own, the intention of creating a 'Greater Somalia' being inscribed both in the state's consitution and on its flag. It had been for that reason, of course, that western governments had declined to supply weapons in the early 1960s. Somali aims, which embraced a territorial claim to a third of neighboring Ethiopia, were pursued by the so-called Western Somalia Liberation Front (WSLF), which had conducted a desultory guerrilla campaign in the

212

Right: Ethiopian troops chanting revolutionary slogans in Addis Ababa while collecting funds for the war against Somalia in February 1978.

Opposite, top: 'Shiftas' surrendering to Kenyan forces in August 1967 during the long-running border conflict of the 1960s between Kenya and these Somali nomads.

Opposite, below: A display of captured Somali equipment including a T-54 tank at an Ethiopian army camp at Diredawa in February 1978.

Below: President Siad Barre of Somalia conferring in June 1978 with President Sadat of Egypt and the veteran Guinean nationalist leader, Sekou Touré.

Ogaden desert region of Ethiopia for some years. There has also been a brief border war with Ethiopia in 1964 while nomads of Somali extraction had been responsible for the border raids into Kenya, known as the 'Shifta War,' between 1963 and 1968.

The region was struck by the Sahel drought in 1974 and this brought to a head dissatisfaction inside Ethiopia with the rule of Emperor Haile Selassie. In September the Emperor was overthrown by a Provisional Military Administrative Committee or 'Derg,' which professed an idealistic socialism. Events thereafter were both confused and bloody as Colonel Mengistu Haile Miriam seized power in a coup in November 1974 and proceeded systematically to eliminate all rivals in the course of the next three years. There were bitter divisions inside Ethiopia, Mengistu being opposed among others by the Marxist group known as the Ethiopian People's Revolutionary Party (EPRP). The secessionist war in the province of Eritrea, ceded to Ethiopia by the British in 1952, also escalated. It had begun in earnest in 1972 when Haile Selassie had abolished Eritrean autonomy and sparked off military action by a variety of predominantly Marxist groups. Ethiopia had been the major recipient of United States military assistance in sub-Saharan Africa but Mengistu's excesses led to the withdrawal of

United States aid in February 1977. The Ethiopians immediately turned to the Soviet Union. The problem was that this posed a difficult dilemma for the Soviets in terms of the existing relationship with Somalia. They had sent some limited assistance to Mengistu in December 1976 and now did so again but both the Soviets and the Cubans were anxious to promote a settlement between two socialist states. The balancing act became increasingly difficult as Barre sought military assistance from the West (which was refused) and sought to exploit the internal disorder in Ethiopia by launching his army, posing as the WSLF, into the Ogaden in July 1977. Ideology had thus failed to transcend local rivalries and, forced to choose, the Soviets leaned toward the apparently more 'progressive' regime in Ethiopia.

By this time Somali forces were enjoying considerable success in the Ogaden, striking at the railway center of Diredawa and the three principal towns of Kebr Dehar, Dagahbur and Jijija. Jijija fell to the Somalis on 14 August 1977 and the Ethiopians also abandoned the strategically important Kara Mandah pass in the Agmar mountains. Having failed to secure Soviet arms, Barre renounced the treaty with the Soviet Union on 13 November 1977 and expelled all Soviet personnel. Just thirteen days later a massive Soviet airlift to Ethiopia began. An estimated 225 Soviet aircraft, mostly Antonov An-22, representing approximately 12 percent of the total Soviet airlift capacity, brought in troops and equipment from the Soviet Union, South Yemen, Angola and Cuba. The equipment brought in by January 1978 included BM-21 rocket launchers, 155mm artillery, 185mm artillery and an estimated 500 tanks while troop levels are estimated to have been 1000 to 1500 Soviet advisers and 11,000 to 15,000 Cubans. The cost was believed to have exceeded $1 billion. It was a considerable logistic achievement, partly used as a training exercise, which fully demonstrated the capacity of the Soviet Union to act at great distances from Russia.

In February 1978 troops with approximately 120 T-54 and T-62 tanks, under the direction of the Soviet General Barisov, spearheaded a counteroffensive against the Somalis. Jijija was recaptured on 6 March and Barre was forced to announce Somali withdrawal, which was completed by 14 March 1978. The attentions of the Soviets then turned to Eritrea between July and November 1978. The Cubans, however, do not appear to have been involved in operations against Eritrea since they had previously lent some support to Eritrean guerrillas and conceivably found Soviet pragmatism distasteful in this respect. Certainly, the Soviets also assisted Mengistu in eliminating the Marxist EPRP and the whole Soviet involvement indicated the rejection of purely ideological considerations for pragmatic expediency. Support for Ethiopia and its territorial integrity was actually guaranteed wider African approval since the OAU had consistently opposed changes in colonial boundaries in the interests of African unity.

Above: Somali troops on parade with their Soviet equipment, soon to be replaced by US equipment in the reversal of roles evoked by the Ogaden war.

Ignoring their considerable investment in Somalia also implied altruism on the part of the Soviets in defending the victim of Somali aggression. In fact, such gains probably outweighed the loss of Somalia's facilities and Ethiopia had equal potential as a strategic base in the long-term.

In the wake of Angola and Ethiopia, to which the West could find little real response beyond castigating Soviet involvement in Africa's 'internal' affairs, it appeared that the Soviet and Cuban influence was widespread throughout the continent. Certainly, the collapse of Portuguese power led to renewed pressure on Ian Smith's government in Rhodesia from both ZAPU and ZANU while opening new infiltration routes for ANC, PAC and SWAPO guerrillas into South Africa and Namibia. Other pro-western governments also appeared to come under pressure, notably Mobutu in Zaire. The Cubans were held responsible for two attempts to topple Mobutu through the invasion of Shaba province (formerly Katanga) by Katanganese gendarmerie based in Angola in April 1977 and May 1978. On the first occasion, Mobutu was saved by the intervention of 1500 Moroccan troops with French and Belgian logistic assistance and, on the second, by French and Belgian troops. In fact, the Cubans denied any knowledge of the first invasion and claimed that they had warned Washington of the imminence of the second and it does appear that Cuban direct involvement was problematical.

Morocco was also threatened in the sense that Cubans, communist Vietnamese and East German advisers were training guerrillas of the Popular Front for the Liberation of Saguiet el Hamra and Rio de Oro (POLISARIO) inside Algeria. POLISARIO claimed to represent the people of the territory known as Western Sahara, the former Spanish Sahara, which Morocco and Mauritania partitioned after Spanish withdrawal in 1976. Mauritania abandoned its claim to the territory in August 1979 but the war continued against the Moroccans. By July 1981 a total of 45 states had recognized POLISARIO's Sahraoui Arab Democratic Republic (SADR) but although Soviet arms certainly found their way to POLISARIO through Algeria and Libya, the Soviet Union itself recognized Moroccan sovereignty and negotiated an agreement for Moroccan phosphates in 1978.

Libyans as well as some Cubans were also involved in propping up the regime of Idi Amin in Uganda, the Libyans becoming involved in the fighting when 4000 Tanzanian troops and 3000 troops of a Uganda National Liberation Front (UNLF) invaded and over-ran Uganda after Amin's continual provocations in November 1978.

In 1978 some 1000 Cubans were reported to be in Mozambique, 400 in Congo-Brazzaville, 300 in Guinea-Bissau, 200 in Equatorial Guinea and 150 in Tanzania while Cubans were also reported as training internal security units in Sierra Leone. Cuban technicians were similarly reported on Saõ Tome and the Cape Verde Islands. There were a reported 6000 East Germans in Mozambique as well as Tanzanians while Nigerians were reported as present in Angola. The Soviet Union meanwhile concluded treaties of friendship with Angola (October 1976), Mozambique (March 1977) and Ethiopia (November 1978). It was reported

Above: Male and female members of the Western Somali Liberation Front on parade in Mogadishu, the invasion of the Ogaden in July 1977 by the Somalis being claimed as the work of the WSLF guerrillas.

in May 1981 that a similar arrangement might have been concluded with Congo-Brazzaville but there is no firm evidence for this. East Germany has also concluded treaties with Angola and Mozambique (both February 1979) and Cuba has formal technical agreements with Mozambique (June 1977) and Tanzania (March 1980). A number of other Eastern Bloc states also have treaties with Angola, Ethiopia and Mozambique and these three may be said to be the principal indigenous Marxist states of sub-Saharan Africa.

The Forces

Ethiopia
The largest armed forces of the three are those of Ethiopia which, in 1983/4, numbered approximately 250,500 men, of whom the great majority of 244,500 were serving in the army (or People's Militia). This force is divided into some 24 infantry divisions (three of which are motorized), four parachute/commando brigades, 30 artillery brigades and 30 air defense brigades. The navy has 2500 men and the air force some 3500 men. In the case of the army, the main armament consists of an estimated 40 American M-47 tanks, 150 T-34 and 700 T-54/55 tanks and 700 artillery pieces of various calibers. The navy has 19 assorted vessels, including one frigate and seven fast attack craft, while the air force has an estimated 107 combat aircraft including 10 MiG-17s, 65 MiG-21s and 20 MiG-23s.

Angola
Angola maintains armed forces of approximately 37,500, of which 35,000 serve in the army (FAPLA), in two motorized infantry brigades, 17 infantry brigades and four anti-aircraft brigades. Armament consists primarily of 17 T-34, 150 T-54 and 50 PT-76 tanks and some 200 artillery pieces. The Angolan navy of 1000 men has 31 vessels, including two fast attack craft, while the 1500 men of the air force fly 67 combat aircraft including 39 MiG-21s, 25 MiG-17Fs and two G-91 R4s.

Mozambique
Mozambique has approximately 12,650 in her armed forces of which 650 serve in the navy, 1000 in the air force and 11,000 in the army (FPLM). The army consists of one tank brigade and seven infantry brigades with 195 T-34 tanks and 250 artillery pieces. The air force has 35 combat aircraft, all MiG-17s, and the navy has 15 patrol craft.

In each of the three countries there are reserves and paramilitary forces amounting to 378,000 in Ethiopia, 10,000 in Angola and 6000 in Mozambique.

Non-African Forces
Much of the real backbone is, however, provided by Cuban or Eastern Bloc advisers. In 1983/4 these were estimated at 12,500 in Ethiopia (11,000 Cubans), 26,150 in Angola (25,000) and a lesser but unknown number in Mozambique. In the case of Ethiopia, however, the total number of Cubans has been reduced during the course of 1984, Castro having announced his intention to withdraw troops in June. It is now believed that as few as 2000 Cubans may remain in Ethiopia. Eastern European advisers are also present in a number of roles in several countries.

Above: A SWAPO guerrilla base captured by South African forces (seen through the window) in Angola. Posters include depictions of Castro, Brezhnev and the Soweto riots.

Right: Members of Angola's Organization of National Defense pictured in May 1976, the movement being a defense and labor corps.

Guerrilla groups in Southern Africa

The principal Soviet or Cuban backed guerrilla groups now operating in Africa are those attempting to infiltrate South Africa and South West Africa/Namibia. The early ANC and PAC attempts at subversion in South Africa failed miserably between 1960 and 1964 but they received an undoubted filip from the collapse of Portuguese Africa, which enabled them to operate from Mozambique, Botswana, Lesotho and Swaziland rather than from the more distant Tanzania as in the past. They also drew new recruits from those fleeing South Africa in the wake of the Soweto riots between June and November 1976 and have re-emerged as a security threat with spectacular sabotage operations, such as that on the SASOL fuel complex in Cape Province in June 1980 and the Koeburg nuclear plant in December 1982. Nevertheless, it cannot be seriously suggested that the highly efficient South African defense forces are over-stretched by PAC or ANC and its South African Communist Party (SACP) allies. The South Africans thus mounted successful raids on ANC/SACP offices and bases in Lesotho in December 1982 and in Mozambique between 1981 and October 1983 while also giving material assistance to the *Resistançia Nacional Mocambicana* (RENAMO), which has fought against the Marxist regime in Mozambique. In the case of South West Africa, SWAPO is generally considered to have a total of approximately 10,000 guerrillas, having opened its campaign of subversion in August 1966 from sanctuaries in Tanzania and Zambia. The withdrawal of the Portuguese enabled SWAPO to operate from southern Angola from 1976 onward but, again, the military threat has been largely contained by the South Africans, whose large scale raids into Angola to destroy SWAPO bases in 1978, 1980, 1981 and especially December 1983 have brought them into contact with Cuban and Angolan forces as well as SWAPO guerrillas. The South Africans' Operation Protea in August-September 1981, for example, killed an estimated 1000 SWAPO, Cuban and Angolan troops as well as four Soviets while a Soviet NCO was captured. Operation Askari in December 1983 killed some 500 SWAPO, Cuban or Angolan troops.

The successes of the South Africans, who have also supported UNITA in its continuing struggle in Angola, have forced South Africa's neighbors to reach an accommodation with Pretoria. A non-aggression pact was concluded with Swaziland in February 1982, leading to the expulsion of the ANC, while Lesotho and Botswana also took steps to clamp down on guerrilla activities by the end of 1983. In February 1984 Angola was forced to negotiate the Lusaka Accords, by which the South Africans offered to withdraw from Angolan territory and agreed to drop support for UNITA in return for the closure of SWAPO sanctuaries in Angola. Similarly, Mozambique concluded the Nkomati agreement with South Africa in March 1984, by which South Africa agreed to cut off assistance to RENAMO in return for the ANC/SACP bases being closed in Mozambique.

Recent political developments

The influence of the Soviet Union and its allies can also be said to have suffered reverses with the continued fall of Soviet- or Cuban-backed African leaders such as Idi Amin in Uganda (1979), the self-styled Emperor Jean Bedel Bokassa in the Central African Republic (1979) and Macias Nguema in Equatorial Guinea (also 1979), none of whose activities improved the image of the Eastern Bloc. Moreover, despite the treaties of friendship with Angola, Ethiopia and Mozambique, none of these could be said to be an entirely quiescent satellite of the Soviets. Angola has leaned toward the West in some respects in search of economic assistance although the continued Cuban presence has proved a major obstacle. In Mozambique, the Soviets appear actually to have been refused naval facilities and, although temporarily eclipsed, the Chinese have regained some influence there. Samora Machel's government has also sought western economic assistance. China has continued to urge African states to heed the example of Somalia and expel the Soviet presence and, of course, it was the group backed by the Chinese – ZANU – that emerged from the war in Rhodesia to form the new government of Zimbabwe upon full legal independence in April 1980. Subsequently, North Koreans have trained the notorious 5th Brigade

of the Zimbabwean army, North Korean advisers also being present in Uganda and the Malagasy Republic. Even Ethiopia, which appears the closest client of the Soviet Union in Africa, has yet to grant naval facilities equal to those lost in Somalia although the Soviets do use Massawa. Morevoer, Mengistu's regime is faced with continuing guerrilla threats in Eritrea from the Eritrean People's Liberation Front (EPLF), in Tigre from the Popular Liberation Front (PLF) and growing unrest among the Oromos of the southwest. Soviet advice and assistance, such as that received in the Ethiopian 'Red Star' offensive in Eritrea in 1982, which was directed by a Soviet General, V I Petkov, has not brought military success while Ethiopia is now estimated to owe the Soviets over $3 billion. Nor can the Soviets apparently offer Mengistu much more than military equipment amid a disastrous drought and famine which has brought a greater response from the West than the Eastern Bloc.

The communist threat

In some respects despite recent reverses, the threat still posed by the Soviet presence in Africa is real. It has at times been exaggerated but, paradoxically, it is also more subtly subversive than sometimes recognized by the general public in western countries. It has been pointed out, for example, that any Soviet attempt to interfere with the sea routes around the Cape of Good Hope, particularly western oil routes, would in itself constitute an act of war. Nevertheless, Soviet naval expansion and the facilities enjoyed by the Soviet Navy in many African ports and anchorages, even if falling short of bases as such, does pose some threat to the ability of the West to retain control of the Indian Ocean and South Atlantic in the event of wider hostilities. Similarly, the domination of southern Africa by the Soviet Union or its allies would imply a *de facto* strategy of denial since the rich mineral resources of the area are absolutely vital to the West. Indeed, the largest deposits of most of the world's key strategic raw materials tend to be located in the Soviet Union and the Republic of South Africa or South West Africa/Namibia.

Yet the influence of the Soviet Union and its allies is very much dependent upon the good will of African governments. On many occasions that good will has evaporated and the Soviets have not been immune to local rivalries which have transcended any ideological considerations on the part of African hosts. Nor is the Soviet Union any longer a new and unknown power in Africa and many African states are clearly suspicious of Soviet opportunism, although the Soviets have contrived to secure a rapport with the majority of the members of the OAU on many individual issues. There have been failures along the way and the Soviets have been forced into protracted military commitments in both Angola and Ethiopia, in the latter apparently at the expense of continuing Cuban support. However, generally speaking, risks taken have paid off handsomely. Above all, the Soviet Union is still well placed with regard to the perennial dilemma of western governments in balancing their strategic interests in Africa, as guaranteed by the continuing existence of the Republic of South Africa as a bastion against communism, against the hostility toward white rule which almost alone unites black Africa.

Above: A March 1984 meeting between Ethiopian leader Mengistu and members of the Soviet government. The late Marshal Ustinov is on Mengistu's right with Chernenko and Gromyko to his left.

Opposite, top: A South African television crew films a meeting between MPLA and South African soldiers forming part of the Joint Monitoring Commission supposedly supervising the withdrawal of forces as part of the South Africa-Angola Lusaka Accords.

Opposite, bottom: Cuban troops training with soldiers of the MPLA at a base in Angola.

7. UNCLE SAM'S BACKYARD

Right: Fidel Castro, doyen of modern Latin American revolutionaries, whose 1959 victory in Cuba inspired numerous other attempts.

Historical background

Latin America has been no stranger to political instability since the majority of its states became independent from Spain or Portugal during the nineteenth century, the pattern of seemingly endemic restlessness persisting after 1945. Social and economic problems have frequently contributed to the outbreak of violence while the usurpation of power by coup d'état has been commonplace. Between 1945 and 1971, for example, there were 81 coups or attempted coups in Latin America.

Yet, the communist world had little interest in Latin America prior to the 1960s, one of the principal means by which the Soviet Union and the People's Republic of China sought to identify with Third World problems – the Afro-Asian People's Solidarity Organization, founded in 1957 – pointedly ignoring Latin America altogether. One reason was that there was little expectation that the largely capitalist states of the continent would prove suitable candidates for revolution and, in the 1920s and 1930s, radical nationalists in Latin America had indeed been more attracted by fascism than communism. Of more importance was the limited capability of the communist powers to influence events in a region that was not only far away from them but in close proximity to the United States. The USA had rarely hesitated to intervene where its interests appeared to be threatened and, under the principles of the Monroe Doctrine of 1823, had sent forces into Latin America on no less than 70 separate occasions between 1830 and 1945. Thus, when communists participated in government in Guatemala in 1953, the Soviets avoided opening diplomatic relations so as not to provoke any US reaction. In the event, the US Central

Intelligence Agency (CIA) backed an exile invasion of Guatemala from Honduras in June 1954 and overthrew the government of Jacobo Arbenz Guzmán when it accepted arms supplies from Czechoslovakia.

Latin American states have also proved suspicious of Soviet intentions and held the communists responsible for the activities of local communist parties such as the attempted risings in El Salvador in 1932 and in Brazil in 1935. Thus, although some Latin American states opened diplomatic relations with the Soviet Union in the 1920s, most severed these in the 1930s. Again, diplomatic relations were opened during World War II but broken off with the beginning of the Cold War so that, by 1953, the Soviets enjoyed diplomatic relations only with Argentina, Mexico and Uruguay. Some fence mending took place in the 1950s but it amounted to little influence for the Soviet Union in the region, a situation not changed until Fidel Castro's victory on Cuba in January 1959.

Cuba and Castro

Castro appears to have had no links with Cuban communists at all prior to July 1958, the local communists having described his attack on the army barracks at Moncada in July 1953 as 'adventurist putschism.' They only began to change their opinion in February 1958 when Castro's forces appeared to be winning the struggle against the government of Fulgencio Batista but, even then, boycotted a general strike called by Castro in April 1958. Nor is there any evidence of Soviet contact with Castro, formal links only being opened with the visit of the Soviet vice premier, A I Mikoyan, to Havana in February 1960. Castro himself had always been connected with the Orthodox

Above: The facade of Moncada Barracks in Cuba's Oriente province showing bullet damage inflicted during Castro's abortive attack on 26 July 1953.

Party, having become involved in student politics at the University of Havana in the late 1940s and having participated in an unsuccessful plot to assassinate the dictator of the Dominican Republic, Rafael Leonidas Trujillo, in 1947. Castro was intending to run for Congress when Batista seized power in a bloodless coup in March 1952. Batista had been something of a social reformer when he had previously ruled in Cuba between 1934 and 1944 but corruption soon overshadowed any further commitment to reform. Castro therefore struck at Moncada with only 111 men in July 1953 but, faced with a 1000 strong garrison, they were forced to retreat and, in the ensuing government operations, Castro was captured. Sentenced to 15 years imprisonment, Castro was released in an amnesty in July 1955 and left for exile in Mexico. His supporters on Cuba, the 26th of July Movement, then planned an urban uprising to coincide with Castro's return with 81 men, including the Argentinian Marxist Ernesto 'Ché' Guevara whom Castro had met in Mexico, in November 1956.

There does not appear to have been any intention of waging a guerrilla war since the rising was expected to succeed but it was crushed before Castro landed from the yacht *Granma* on 2 December. Troops were encountered three days later and only 22 survivors were reunited in the rugged Sierra Maestra mountains of the Oriente province of eastern Cuba.

Thus the guerrilla campaign resulted from a lack of any practical alternative rather than from preconceived strategy. Moreover, in developing the so-called 'foco' theory of guerrilla warfare from the Cuban experience, Guevara and the French Marxist, Régis Debray, chose to emphasize the rural nature of the ensuing struggle at the expense of the true facts. For Castro's guerrillas simply could not have survived without the assistance of the 26th of July Movement in the urban centers, such as the network established by Frank Pais in Santiago de Cuba. Urban action, though invariably a failure, also contributed to the tying down of an estimated 15,000 government troops, a process in which other groups such as the *Directorio Revolucionario* were involved as well as Castro's own supporters.

The course of the war also owed much to the sheer incompetence of Batista's army, which despite expanding to a total of 38,000 men by June 1958, never succeeded in eliminating a guerrilla threat which, according to Castro, consisted of only 1500 men in January 1959 and had been as few as 100 in April 1957. It says much for the morale of Batista's army that only 200 government troops are estimated to have been killed in the entire campaign between 1956 and 1959. Batista also lost the support of most of Cuban society so that, in a sense, Castro hardly played a role at all in creating conditions favorable to a revolution. The United States also cut off military supplies to Batista in March 1958 and, in the face of the guerrilla advance in that year, the Batista regime simply crumbled. The provincial capital of Las Villas province, Santa Clara, fell to the guerrillas on 28 December 1958 and Batista fled into exile on 1 January 1959, Castro arriving in Havana to claim his victory on 8 January 1959.

Ostensibly, Castro had fought to restore the 1940 constitution, a radical document for its time but far from communist. Initially, too, he was only a member of the coalition government that was installed and he even visited the United States in April 1959. But President Dwight D Eisenhower's administration was intensely suspicious of the likely direction of Cuban politics and in July 1960, following Mikoyan's visit, the US cut off its quota of sugar purchases from Cuba. With the exception of Mexico, all the other members of the Organization of American States (OAS), founded in 1948, also complied with US policy in imposing economic sanctions and suspending diplomatic relations. Cuba, however, was not formally ejected from the OAS although membership was suspended in 1962. Curiously, the Cubans have also never

Above: Cuban troops and militia with a launch captured from the CIA-backed exiles who landed at the Bay of Pigs in May 1961.

Far left, bottom: Ernesto 'Ché' Guevara driving a tractor while Cuba's Minister of Industry, a post to which he was appointed in 1961.

Far left, top: Castro meeting the then US Vice-President, Richard Nixon, in April 1959 before the American administration took exception to the new Cuban regime.

Above: Members of Cuba's Revolutionary Court, who tried 1179 persons implicated in the Bay of Pigs affair, in Havana in April 1962.

Above right: US aerial reconnaissance photograph of Soviet missiles at Port Mariel on Cuba on 4 November 1962, shortly after the conclusion of the Cuban Missile Crisis.

Below right: US aerial photograph of Soviet *Komar* class missile patrol boats at Banes, Cuba on 3 November 1962.

formally abrogated their adherence to the 1947 Rio Treaty of hemispheric defense, which makes Castro technically an ally of the United States. A further paradox, which persists, is the existence of a US base at Guantanamo on Cuba's eastern coast under a treaty of 1934.

Initially, Castro had purchased some military equipment from western Europe but by the end of 1960 some 3000 Soviet and Czech advisers had arrived and both the Soviet Union and China had stepped in to buy Cuban sugar. Nikita Khruschchev, the Soviet leader, had gone further in a speech on 9 July 1960 in which he stated, 'Figuratively speaking, in case of need, Soviet artillerymen can support the Cuban people with their rocket fire should the aggressive forces of the Pentagon dare to start intervention against Cuba.' Castro, who claimed that the US was threatening Cuba's independence, welcomed Soviet support in his 'First Declaration of Havana' in September 1960 but it still appears that the Soviets had no expectation at this stage that Castro was more than a radical nationalist. Castro only proclaimed himself a Marxist in July 1961, significantly after the attempted Bay of Pigs invasion of Cuba, and was not to create a unified communist party on the island until October 1965. Soviet support in 1960 should also be seen in the context of the deteriorating relationship between the superpowers after the U-2 incident of May 1960

The Bay of Pigs fiasco, in which 1300 ill-trained Cuban exiles were landed back on the island on 17 April 1961, had been contemplated by the CIA as early as January 1960 although it did not gain Eisenhower's approval until eight months later. President John F Kennedy, who took office in January 1961, had misgivings and reduced the air cover to six B-26 bombers while directing that US personnel should not become directly involved. Without sufficient air cover, the invaders were totally destroyed, 120 being

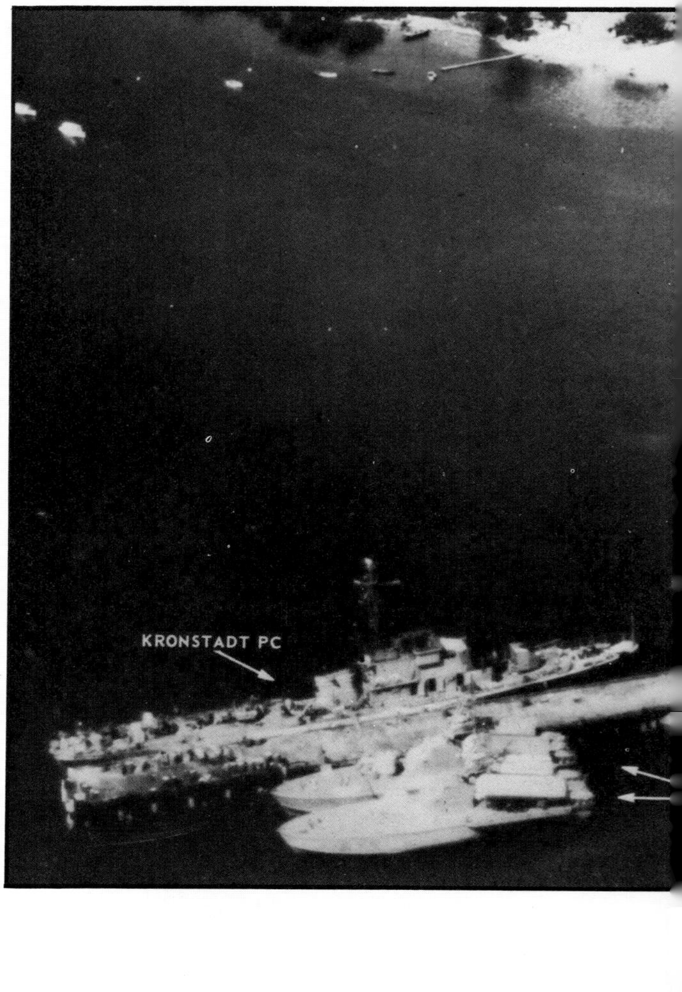

Above: Three US reconnaissance pilots posing in front of a McDonnell RF-101 aircraft, such unarmed planes playing a vital role in exposing the Soviet missile build up on Cuba.

killed and the rest captured. After the Bay of Pigs there could be no accommodation between Castro and the Americans and, inexorably, Cuba was drawn into the wider confrontation between Khrushchev and a US president whom the Soviet leader clearly believed he could defeat on the political stage. It would appear that the Soviet decision to exploit the proximity of Cuba to the United States by placing missiles there was taken in June 1962, the Soviets claiming that Castro had requested additional military support although Castro always denied this. The Soviet gamble to arrest the decline in their strategic nuclear posture vis à vis the United States was then played out in October 1962 without reference to Castro. As is well known, the Kennedy administration chose to impose a naval blockade to prevent more missiles reaching Cuba, their presence having been detected on 14 October and announced to the world on 22 October. The Americans had been careful to leave Khrushchev some room to maneuver and, on 27 October, the Soviets agreed to remove 42 missiles already sent to Cuba in return for a US guarantee not to invade. Soviet Il-28 bombers were also withdrawn.

A distinct cooling in relations between Moscow and Havana followed the Cuban Missile Crisis, in which Castro felt he had been humiliated, and the Cubans briefly flirted with the Chinese communists. At home Castro consistently ignored Soviet advice on the transformation of the Cuban economy and offered encouragement elsewhere in Latin America to groups that the Soviets regarded as leftist extremists rather than to orthodox communist parties. In the 1960s with the ideological differences between the Soviet Union and China becoming more apparent, Castro added yet a further dimension to the multiplicity of left-wing parties in Latin America, some 20 parties taking a pro-Moscow line, 11 a pro-Peking line and at least five a pro-Havana line. Indeed, there is some evidence of Soviet involvement with more orthodox communists in Cuba, Anibal Escalante being exiled in 1962 by Castro and subsequently arrested in 1968, after his return, for attempting to enlist Soviet support for a 'mico-faction' opposed to Castro.

Nor did the Soviets and Castro agree on the export of Cuban-style revolution to the remainder of Latin America. Despite Khrushchev's advocacy of 'wars of national liberation' in January 1961, to which rhetoric Castro had added his 'Second Declaration of Havana' in February 1962, Khrushchev still essentially believed in 'peaceful co-existence' between the superpowers. The advantage, as far as the Soviets interpreted this, would be that the favorable climate they detected in the Third World generally toward socialism would result in the emergence of pro-Soviet governments, free of the risk of United States intervention. Khrushchev's successors were also keen to establish diplomatic relations with Latin American states. Thus Castro's proclaimed revolutionary intentions were an embarrassment, especially as the Soviets regarded such action as doomed to failure in Latin America. Nevertheless, for fear of Chinese claims that they were 'revisionists,' the Soviets paid lip service to Castro's 'left adventurism.' They accepted the ringing declarations for revolution that Castro extracted frpm a conference of Latin American communist parties at Havana in November 1964, which endorsed revolutionary struggle in Venezuela, Colombia, Honduras, Paraguay and Haiti; those at the Tricontinental conference at Havana in January 1966, at which Castro called for 'two, three, more Vietnams'; and in Castro's newly-sponsored Organization for Latin American Solidarity in August 1967.

Theories of guerrilla warfare

The problem with revolutionary guerrilla struggle was, however, that it was doomed to failure because the rest of Latin America was not, as the Cubans assumed, a carbon copy of Cuba between 1956 and 1959. Guevara and Debray believed that a revolutionary struggle could be waged and won entirely in the countryside rather than the urban areas, whose support in the Cuban campaign they had conveniently forgotten. They also believed that guerrillas could beat the armed forces of a state and, even more problematically, they assumed that there was no need for a lengthy period of political preparation prior to the commencement of insurgency as in the classic Maoist approach to rural guerrilla warfare. They believed, indeed, that there would always be a sufficient minimum level of popular discontent with any government to provide the basis of popular support for the guerrillas. The guerrillas themselves would, in any case, be capable of generating their own success through direct action in a series of hard-hitting highly mobile operations. The failure of the government to suppress this small elite band of guerrillas would force it into repression against the population as a whole thus exposing the corrupt nature of the system to the people. The guerrillas would then have provided, through a 'band-wagon effect' of 'revolution within the revolution,' the 'foco' or focus of wider insurrection.

Such assumptions rested on fundamental misinterpretation of events on Cuba. Other Latin American states did not face pressing issues of land reform. In Bolivia, where Guevara was to meet his death, land reform had already provided the peasantry with more land than they could use efficiently while the drift of population to the towns had relieved any pressure on the land. Mountain Indians in the Andes through their primitive nature, illiteracy and superstition, deference to authority, and low expectations were singularly unlikely candidates for revolution. This was all the more so when the elite band of guerrillas were also likely to fall foul of sheer xenophobia in such mountain communities – in Bolivia, the Indians helped track down Guevara's assorted band of Cubans, Argentinians, Peruvians and Bolivians who had not even mastered local Indian languages. In urban areas, which Guevara and Debray scorned, workers facing unemployment or under-employment were also cautious to an extreme of jeopardizing their employment through subversive activities but, whereas Batista had alienated all sectors of society, other Latin American societies were generally less volatile than that on Cuba and overwhelm-

ingly favored the status quo. Through ideological differences, local communist parties owing allegiance to Moscow had no interest in direct action and even the Maoist groups were not apparently prepared to put into practice the revolution they preached.

Nor was the rest of Latin America like Cuba in terms of terrain. The Sierra Maestra mountains had been difficult but sufficiently broken to allow mobility and sufficiently vegetated to provide cover. In Latin America jungles afforded cover only at the expense of mobility, Guevara's band spending most of its time in Bolivia either lost or painfully hacking through terrain that was the very antithesis of that required for the highly mobile operations 'foco' demanded. Guevara and Debray were also at error in believing that guerrillas could defeat armed forces almost as a matter of course for Latin American armies were far more formidable than that of Batista. Armed forces were generally more popular with the population of such states than the police and had a tradition of assisting civil development so that, being also conscripted forces in which peasants often assisted development projects in their own localities, they were far more representative of society than the guerrillas.

American involvement

An even more important difference between Cuba and what occurred in Latin America in the 1960s was that the United States was no longer prepared to remain inactive. The Kennedy

Left: Ché Guevara (second left) in the Bolivian jungle during his disastrous attempt to export Cuban style 'foco' guerrilla warfare to Latin America.

Below: Police arresting a demonstrator at Barquisimeto in Venezuela during President Romulo Betancourt's successful campaign against FARN guerrillas.

Right: A Venezuelan policeman standing by the body of an intelligence agent killed by guerrillas in Caracas in December 1963.

administration had responded to Khrushchev's call for wars of national liberation with the 'Alliance for Progress' in March 1961, military advice being combined with large injections of economic aid. States felt to be particularly at risk such as Venezuela, Peru and Guatemala received US military missions while the 'Green Berets' of the US 8th Special Forces Group, based on Fort Gulick in the Panama Canal Zone, became directly involved in operations. In 1965, for example, 52 anti-subversive operations were carried out and probably in excess of 400 between 1962 and 1968. Over 1000 Green Berets were in Guatemala after 1966 while it was a US-trained Ranger battalion of the Bolivian army that tracked down Guevara in October 1967. As many as 20,000 Latin Americans received military training in the Canal Zone between 1962 and 1970 while a further 3500 Latin American police officers were trained by the US Agency for International Development in Washington and Panama between 1966 and 1970. Financial assistance to Ecuador alone amounted to $1.5 million in 1962 for improvements in such areas as sanitation, health care, water supplies and communications. The United States was also prepared to intervene directly as was proved by the case of the Dominican Republic in May 1965, where 21,000 US Marines were landed to prevent the leftist Juan Bosch returning to power. An OAS force remained on the island until September 1966.

Castro's renewed Soviet ties

Thus 'foco' failed wherever it was attempted – in Columbia in 1961, in Guatemala and Ecuador in 1963, and in Peru in 1965. Most significantly, it failed in Bolivia, Guevara being trapped and killed in October 1967. A similar failure awaited an urban-based insurgency in Venezuela between 1962 and 1965. Such failures, and notably Guevara's death, undoubtedly had an impact upon Castro who was finding it increasingly difficult to balance Moscow against Peking. There were still Soviet policies of which he disapproved but Cuba's economic problems were also pressing and, in August 1968, Castro gave verbal support to the Soviet intervention in Czechoslovakia. In the absence of any other source of economic assistance, Castro's promotion of revolution having cut him off from the possibility of western aid, Cuba rapidly became totally dependent upon the Soviet Union. The Soviets consciously used the economic leverage they possessed, especially in the manipulation of oil supplies, and Cuba joined the Council for Mutual Economic Assistance (COMECON) in July 1972. Castro was concerned to modernize his armed forces and ensure greater protection, the Soviets responding to this need with their first naval visit in July 1969 although the Soviets subsequently agreed in October 1970 not to develop Cienfuegos as a submarine base when the United States objected to the possible violation of the 1962 agreement. Castro made visits to the Soviet Union in 1972 and championed the Soviets within the Non-Aligned Movement, the seal being set on the Cuban-Soviet rapprochement by the visit of Leonid Brezhnev to Cuba in January 1974.

Urban guerrillas

The Soviets appear to have extracted an agreement from Castro to stop exporting revolution to the rest of Latin America for they firmly believed that there was far more promise in investing in the peaceful transition to revolution. The Soviets doubted the success of the ultra-left revolutionary groups that were now emerging in many Latin American states and indulging in urban guerrilla warfare. They believed that this merely provided reactionary forces with greater opportunities for stifling revolution and distrusted leadership of the revolution by armed groups rather than a Marxist-Leninist vanguard party. To a large extent they were correct in their estimation of the chances of the urban guerrillas, urban guerrilla theory as advocated by Carlos Marighela in Brazil being curiously like 'foco.' Marighela believed that society and the governmental apparatus could be paralyzed by the action of a small elite band in the cities. It owed much to the recognition that the great majority of the population of Latin America now lived in urban areas where a combination of inflation, unemployment and a relatively youthful age structure created conditions for dissatisfaction in the slums and shanties surrounding larger cities. It was assumed that 'armed propaganda' would provoke a general repression which, like the reaction to 'foco' in the countryside would induce wider insurrection. In practice, the response of Latin American governments faced with urban insurgency was so severe that it quickly destroyed the guerrillas and rendered them incapable of exploiting any popular unrest that developed as a result. Thus maximum force and institutionalized counterterror such as in Argentina's *Guerra sucia* or 'dirty war' after the Argentinian army's takeover in March 1976, resulted in the disappearance of most of the urban guerrilla groups by the mid 1970s. In any event, such urban guerrilla groups as Marighela's *Ação Liberatadora Nacional* (ALN) in Brazil and the Trotskyist *Ejército Revolucionario del Pueblo* (ERP) in Argentina were most often those that had broken away from more orthodox communist parties and were too diverse in aims and ideology to attract even Castro's support let alone that of the Soviet Union. Certainly, the Soviets forced Castro to drop his support for the *Fuerzas Armadas de Liberación Nacional* (FALN) in Venezuela although they may have had links with the *Tupamaros* in Uruguay.

The development of socialism

A second reason for advocating peaceful transition to revolution was that the Soviets perceived the need to develop relations with states such as Venezuela, an oil producer, and states such as Brazil and Argentina that could remedy some of Moscow's shortfalls in grain

production. But a third, and more important, reason was that there appeared to be encouraging signs that alternative means to violence were bringing results. In 1969, for example, the so-called Andean Pact of Chile, Peru, Bolivia, Colombia and Ecuador aimed at economic co-operation that would exclude foreign, primarily US, capital. The Catholic Church was becoming more radical in Latin America and many states were experiencing the beginnings of broad left united fronts in the political arena which held out the hope of power through the ballot box. Even military regimes did not seem to be beyond hope. In Peru in October 1968 a military coup brought to power a leftist military government headed by General Juan Velasco Alvarado which was anti-American and intent on transforming the Peruvian economy. Nationalization policies brought conflict with the United States and, in 1973, Peru became the first state in the western hermisphere outside Cuba to accept Soviet arms, some 200 T-55 tanks being delivered in November. Similarly, there was a left-wing military coup in Bolivia in July 1970 led by General Juan José Torres. Torres did not survive for long, being ousted in 1971, while Velasco was to be overthrown in August 1975 and left-wing elements purged by the time the government was finally returned to

civilian hands in July 1980. Nevertheless, at the time they came to power, Velasco and Torres represented the possibility of more radical nationalists emerging while the Soviets were not unduly deterred by ideology from dealing even with right-wing military governments.

An even greater example of the apparent success in investing in peaceful transition appeared to be the emergence of the socialist coalition government of Salvador Allende in Chile in September 1970. The Chilean Communist Party had played a prominent role in forging the six-party Popular Unity Front, which formed a minority government on 36 percent of the popular vote, but Allende himself was leader of the Socialist Party and claimed to be a populist democrat who wanted socialism without totalitarianism. However, the pace of nationalization aroused opposition and the Chilean revolution was rocked by a fall in copper prices and a staggering 300 percent rate of inflation. The Soviets were remarkably cautious in their approach to Allende's regime and did not offer as much economic assistance as was demanded, conceivably because they were mindful of the cost of keeping Cuba afloat (an estimated $7 billion between 1960 and 1973). In any case, their economic advice was ignored by Allende and the Chilean armed forces turned down the Soviet arms offered them in 1971. For all their caution, this did not lead the Soviets to prevent Castro investing in a diplomatic sense in the Allende regime although they appear to have disliked Cuban and North Korean use of Chile as a subversive base in Latin America, 150 Cubans being expelled from Chile when Allende fell. That collapse was preceded by strikes by the professional classes and by transport workers in August 1973 and by censure of the government by the Chilean Congress, which invited military intervention. With the active support of the CIA, the Chilean army overthrew Allende on 11 September 1973, the new government being headed by General Augusto Pinochet Ugarte. The course of events was a blow to the Soviets and they indulged in some furious propaganda against the Pinochet government, although a Soviet dissident was subsequently swopped for the Chilean communist, Luis Corvalan, in 1976. The Chinese, who had also been recognized by Allende's government, argued that the fall of Allende had been entirely due to Soviet influence over the course of the Chilean revolution and continued to enjoy diplomatic relations with the Pinochet government.

The failure of both the peaceful and the revolutionary paths to revolution may well have had an important influence in Castro's willingness to shift his vision beyond Latin America. The Third World generally was of more importance to Castro than it was to the Soviets and, in some respects, Cuban intervention in Angola in 1975 may well have been more a case of the Cubans involving the Soviets than vice versa. It was popular with Cuba's black population and with radical governments in the Caribbean such as that of Michael Manley on Jamaica and Forbes Burnham in Guyana, the latter country providing one of the staging posts for Cuban flights to Angola in 1975 and to Ethiopia in 1977. It also improved Castro's standing in the Non-Aligned Movement, which he aspired to lead, although this was later effectively sabotaged by Soviet intervention in Afghanistan just four months after Castro had assumed the chairmanship at the movement's conference in Havana in September 1979.

At the same time, through Cuba's lessening support for revolutionary groups, Cuba's standing improved in Latin America. In November 1974 the OAS voted by twelve votes to three with six abstentions to lift the economic sanctions imposed in the 1960s. This was less than the required two-thirds majority but those states that had voted for Cuba indicated that they would now lift sanctions. Cuba gained some kudos from sending medical teams to assist in natural disasters in Nicaragua in 1972 and in Honduras in 1974; she joined 24 other Latin American and Caribbean states in the Latin American Economic System in October 1975 and received the diplomatic recognition of the Carter administration in the United States in 1977. In the following year, Venezuela supplied Cuba with oil. But this apparent 're-integration' into Latin America did not necessarily imply that Castro had lost all interest in promoting revolution and supporting 'progressive' regimes and new tensions arose in both the Caribbean and Central America in 1979.

Far left, top: The ruins of the Presidential Palace in Santiago, Chile after the army's coup against the Allende regime in September 1973.

Far left, bottom: Chilean troops clashing with demonstrators. Military victory against the urban guerrillas of Latin America has frequently been at the expense of democracy.

Left: The Chilean President, Salvador Allende, who committed suicide during the Chilean army's coup against his left-wing regime in September 1973.

Top: Part of the Soviet weaponry captured by US forces when they invaded Grenada in October 1983.

Top right: A US Marine displaying crates of weapons captured on Grenada, which the Cubans claimed had been supplied only for local use by the militia.

Above: A paratrooper of the US 82nd Airborne Division with American students of the St Georges Medical School, whose safety was one justification used for US intervention on Grenada.

Right: Castro welcomes back Cubans wounded in the fighting on Grenada to Havana on 2 November 1983. In all, 784 Cubans were on the Caribbean island.

Recent developments

Grenada

Both Central America and the Caribbean are of deep significance to the United States since well over 50 percent of its trade passes through the waters of the region, the Panama Canal remaining a vital shipping link. Moreover, some two-thirds of the United States' oil supplies are now derived from such states in the region as Mexico and Venezuela. Thus, the United States was bound to view with alarm the development of crises in the course of 1979. In the Caribbean, the focus of attention was the small island of Grenada, just 133 square miles in extent but only 90 miles from Venezuela. Not untypical of the 'micro-states' that had come to independence in the Caribbean in the 1970s, Grenada was governed by Eric Gairy's United Labor Party after independence in February 1974. Given to lecturing the United Nations on the necessity for investigating unidentified flying objects, Gairy had contacts with the Pinochet government of Chile and also maintained his autocratic style of government with the assistance of a band of thugs known as the 'Mongoose Gang.' However, on 13 March 1979, while Gairy was again at the UN, the opposition New Jewel Movement (NJM) led by Maurice Bishop seized power in a bloodless coup. A self-proclaimed populist 'revolutionary-democratic'

Left: Daniel Ortega, the leader of the Sandinista government that took power in Nicaragua in July 1979.

Below: A Soviet-supplied ZSU-23-4 23mm anti-aircraft cannon of the Cuban *Fuerzas Armadas Revolucionarias*.

Right: Edén Pastora, the former 'Commander Zero' of the Sandinista army in Nicaragua, pictured in April 1983 while leading the anti-Sandinista Contra group known as ARDE.

Far right, top: A US reconnaissance photograph of Sandino airfield in Nicaragua taken in March 1981 and released in January 1982 amid the persistent rumors of the receipt of MiG fighters by the Sandinista regime.

Far right, bottom: Weapons training for women workers in a Nicaraguan factory during 1984, widespread military training having been introduced by the Sandinistas.

party, the NJM did not call itself communist but was the only non-communist government to vote with the Soviet Union against the censure motion on the Soviets for their intervention in Afghanistan in the UN General Assembly in January 1980.

Grenada also swiftly received Cuban assistance in the form of building materials and, allegedly, weapons as early as March 1979. Cuban medical teams arrived in June and the Cubans appear to have begun discussion of the construction of a new airfield at Point Salines, which the NJM saw as an important development project, in August. Bishop went on to sign various agreements with both Cuba and the Soviet Union on military training for Grenada's armed forces. The establishment of the Bishop government appeared to have a 'knock-on' effect with violence occurring on Dominica's Union Island, where order was restored by troops from Barbados; a left-wing turn in St Lucia; and the emergence of similar radical movements in Dominica and Antigua. In November 1979 Bishop announced that the Cubans would provide $9.2 million worth of aid and 250 advisers. These apparent upheavals in the Caribbean taken together with the existence of radical governments in both Jamaica and Guyana was of considerable concern to the United States.

Nicaragua

Further alarm bells were rung through events in Central America, since in July 1979 the last of a dynasty of *caudillos*, Anastasio Somoza, fled Nicaragua in face of the culmination of a successful guerrilla campaign by the *Frente Sandinista de Liberación Nacional* (FSLN). Named after a Nicaraguan general, César Augusto Sandino, who had fought against United States intervention in Nicaragua between 1928 and 1933 and who had been murdered by the first Somoza in 1934, the Sandinistas had been formed as long ago as 1958. The guerrilla campaign did not, however, gain any momentum until 1974 and received new impetus from the decision of the Carter administration to cut off military supplies to Somoza in 1977 because of Somoza's lamentable record on human rights. The Sandinistas also enjoyed more popular support from all classes of society than any other revolutionary group in Latin America, Somoza alienating virtually all his likely supporters, not least by appropriating world relief funds poured into the country after the Managua earthquake in 1972. Cuba appears to have played little role in the success of the Sandinistas, although they are credited with preserving the unity of guerrilla factions during 1978, and the Carter administration appears to have adopted an ambivalent rather than an overtly hostile attitude to the new government initially. But more radical elements within the FSLN emerged in 1980 and 1981 and, by February 1982, there were a reported 1800 to 2000 Cuban and 50 Soviet advisers in Nicaragua. By no means all of these were military advisers and it is at least debatable that the build up of Cuban assistance to Nicaragua reflected the growing hostility of the United States to the Sandinistas.

There were those in Washington who believed that Marxist influence would soon prevail in Nicaragua and that there was an inherent danger that a foreign power might be able to establish the first hostile military facilities on the mainland of Central America, resulting in a kind of 'domino' effect which would disturb the stability of all Nicaragua's neighbors and, ultimately, Mexico.

El Salvador

Certainly, events in Nicaragua gave immense encouragement to guerrillas in the neighboring state of El Salvador. The situation within El Salvador resembled more of a civil war than a revolutionary struggle against the government, which was dominated by the military and the 'oligarchy.' Equally, there was a struggle within the ruling group, the presidential elections of both 1972 and 1977 being blatantly rigged against the Christian Democrats.

Recognizing the lessons of Somoza's fall, more progressive military officers seized power in October 1979 and overthrew the government of General Carlos Humberto Romero but attempts at reform were to be continually frustrated. Amid the increasing polarization of the population, a number of revolutionary groups came together to form *Frente Farabundo Martí para la Liberación Nacional* (FMLN) in May 1980. The guerrilla coalition takes its name

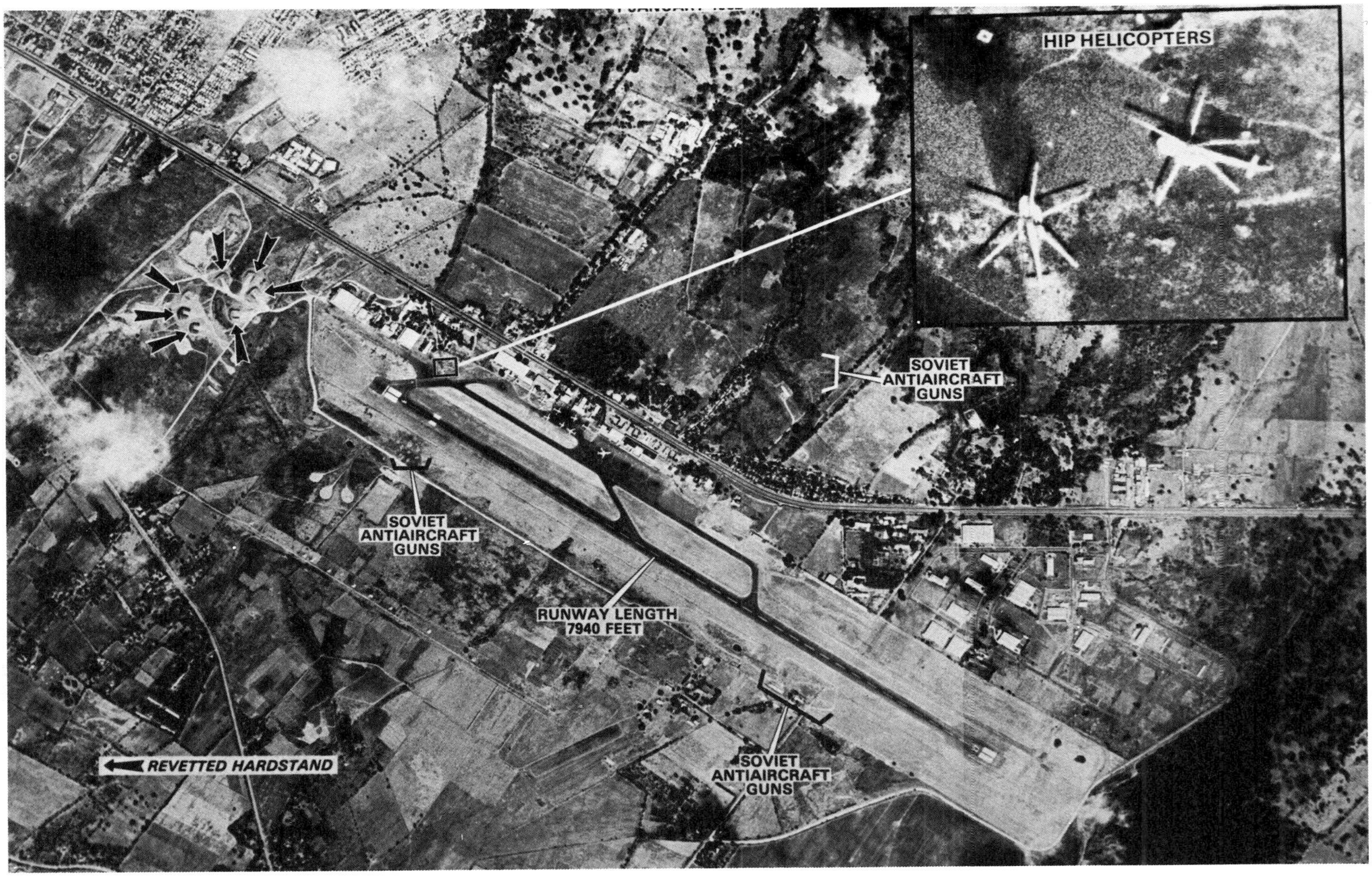

from Farabundo Martí, a prominent communist executed in January 1932 after the failure of a brief insurrection. It was assumed that the FMLN was receiving assistance from the Sandinistas after the fall of Somoza in Nicaragua but the Sandinistas denied it. Events in both Nicaragua and El Salvador also appear to have contributed to outbreaks of violence in Honduras in October 1980 and Costa Rica in March 1981 and, of course, the simmering guerrilla war in Guatemala has cost at least 22,000 lives since 1979 alone, having begun in the 1960s.

US Policy

For those in Washington disposed to believe in a communist conspiracy, the final confirmation in 1979 was the Carter administration's 'discovery' of a Soviet 'combat brigade' of 2600 men with 40 tanks and 60 armored personnel carriers on Cuba in September 1979. The Cubans and the Soviets strenuously denied that there were any more Soviet troops on the island than there had been since 1962 and insisted that they continued to adhere to the principles of the Kennedy/Khrushchev agreement of 1962. In the event, more immediate problems confronted the United States in Iran but Carter announced the establishment of a Task Force at Key West in Florida in October and 'window dressing' maneuvers were held off Guantanamo. Carter also resumed military aid to El Salvador, cut off in 1977 like that to Nicaragua over the lack of human rights in the country. Others also clearly

Above: A poster of Ché Guevara looks down on a parade of Cuban missile units in Havana.

made something of a connection between events in the region and Cuba with dire consequences for the new relationships Castro had so painfully forged in the 1970s. Colombia and Costa Rica both cut off diplomatic relations in 1981, the Colombians alleging that Cubans were assisting guerrillas who had attempted a seaborne invasion of the state. Venezuela cut off oil supplies in April 1980, when relations were strained during the mass exodus of Cubans from Cuba, and, with the electoral defeat of Michael Manley on Jamaica in November 1980, that island too moved toward cessation of relations by October 1981.

Cuba's isolation was increased by the advent of the Reagan administration in the United States which viewed developments in the Caribbean and Central America as sinister. Aid to El Salvador was substantially increased with the small number of 19 US advisers of 1979 rising rapidly to 56, the first being subsequently killed in El Salvador in May 1983. Four specialized *cadoza* or 'hunter' anti-guerrilla battalions were trained and between 1981 and 1983 US aid totalled some $235 million. Despite the caution of the US Congress, aid in 1984 to date has reached $195 million. Aid was also poured into El Salvador's neighbors, Guatemala receiving some $44 million in 1983, Costa Rica some $200 million and Honduras some $1 billion since 1980. The United States has also encouraged El Salvador and Honduras to patch up their outstanding differences, which resulted in the 'Soccer War' in 1969, with a treaty in November 1980. The Central American Defense Council (CONDECA), formed originally in 1963 but defunct since the Soccer War, was also revived in late 1983 with Costa Rica, El Salvador, Honduras, Venezuela, Colombia and the United States as members.

United States' investment in such states as Honduras and Costa Rica, which abolished its own armed forces in 1948, is also specifically aimed to destabilize Nicaragua while stabilizing the former countries. A positive decision to undermine the Sandinistas appears to have been taken in November 1981, the CIA organizing counter-revolutionary forces or 'Contras.' These comprise former Somoza supporters in the *Fuerza Democrática Nicaraguense* (FDN), operating from Honduras, and the *Alianza Revolucionaria Democrática* (ARDE), operating from Costa Rica and comprising former Sandinistas such as Edén Pastora, who opposed the leftward turn in Nicaragua. The FDN is thought to number some 7000 and ARDE some 5000 men. The CIA is also regarded as having been responsible for attacks on Nicaraguan ports and oil storage facilities by aircraft and launches in September 1983 and, most recently, for the mining of Nicaraguan ports in April 1984. United States forces have also continued to hold large scale maneuvers in neighboring Honduras, the United States' activities being

Left: The same Guevara poster and infantry of the most powerful armed forces in Central America – Cuba's FAR.

Below left: Castro with the Angolan leader, Dr Agostinho Neto, to whose assistance Cuba committed troops in 1975 during the Angolan civil war. They still remain there.

likened by opponents to a 'slow motion Bay of Pigs.'

Nor did the Reagan administration neglect the Caribbean, the so-called Caribbean Basin Initiative being announced in February 1982 to increase US economic assistance to the region. The Reagan administration was also hostile to the Bishop government on Grenada, despite its continuing membership of the Caribbean economic community (CARICOM) and its participation in the Organization of East Caribbean States (OECS) in June 1981. Bishop had, however, established close relations with the Sandinista government of Nicaragua and organized a First International Conference in Solidarity with Grenada in November 1981 at which 41 states were represented. There was also the continuing Cuban presence and the construction of the 9800ft runway at Point Salines, on which the Cubans had spent over $60 million. Events were precipitated by the unexpected arrest of Bishop by his Marxist deputy, Bernard Coard, on 13 October 1983 as the culmination of an internal power struggle to control the direction of the Grenadian revolution. Bishop was undoubtedly popular and was freed from house arrest by crowds on 19 October, only to be shot dead with four other members of his Cabinet and numerous civilians by Grenadian troops. A Revolutionary Military Council was announced under the leadership of the army commander, Hudson Austin, though Coard was the dominant figure. Concern at Bishop's arrest had already been voiced by Barbados as well as members of the OECS, Barbados and the OECS agreeing to act on 21 October 1983. Jamaica also joined the planning of the contemplated military action and the United States was invited to participate on the same day.

It was subsequently claimed that a United States and OECS intervention was justified both by the OECS treaty and by an invitation to act extended to the OECS by Grenada's Governor-General, Sir Paul Scoon, but Scoon's precise role is open to doubt and it seems unlikely that the United States did not already possess contingency plans for such an eventuality. Intervention was also justified by the construction of the Points Stalines airfield, which was portrayed as a military installation, and by the concern for the safety of over 1000 US citizens, mostly medical students at the island's St George's University Medical School. On 25 October, therefore, 'Operation Urgent Fury' saw some 400 Marines of US 22nd Marine Amphibious Unit landing to secure the civil airfield at Pearls while members of an elite Marine 'Seal' unit moved to rescue Scoon at Government House. Some 500 Rangers were landed by Hercules aircraft at Point Salines with 5000 men of US 82nd Airborne Division flown in from North Carolina once the runway was secured. Some 300 troops and police from Barbados and Jamaica and the OECS member states of Antigua, Dominica, St Lucia and St Vincent were also brought in once the landing areas were secured and to guard prisoners.

Although the United States had accused the Bishop government of having up to 1600 Cubans on Grenada, they evidently believed that there were no more than 300 and that they would not fight. Castro had indeed refused Austin's request to send troops on 22 October but had ordered Cubans to defend themselves. In all 784 Cubans were present, 636 being construction workers and only 25 military advisers as such but they put up a stiff fight and the island was not fully secured until 28 October. There was little resistance from the Grenada militia, which was largely loyal to Bishop and would not come out to defend Coard's regime, but the 1000 strong army did resist. In the fighting 24 Cubans, 16 Grenadians and 18 US personnel were killed. A large number of weapons were recovered but these were of a very mixed kind, the Cubans claiming that they were for the popular militia and had been supplied on the understanding that they were not exported elsewhere. Others besides Cubans were also apprehended including East Germans in the telephone exchange and 32 Soviets. However, for all its rhetorical contribution to anti-American propaganda in the region, not least through a powerful radio transmitter, Grenada did not appear to be the 'Soviet-Cuban colony being readied as a major military bastion to export terror and undermine democracy' that President Reagan claimed and, outside the United States, world reaction to the American intervention was less than wholeheartedly favorable.

Communist Forces

Cuba

The Cuban Armed Forces, known as the *Fuerzas Armadas Revolucionarias* (FAR) are the most powerful in Central America. Numbering some 153,000 in 1984, some 125,000 serve in the army with 12,000 in the navy and 16,000 in the air force. The Cuban army has an armored division, three mechanized divisions, six infantry divisions and an amphibious assault brigade while its main armament consists of 1200 artillery pieces of assorted calibers and 660 main battle tanks. The latter include 60 T-62s, 250 T-54/55s and 350 T-34s. The navy has 103 vessels but this includes only one frigate and three submarines as larger units and the air force has 250 combat aircraft, including 15 MiG-17s, 30 MiG-21s and 51 MiG-23s of different versions. Paramilitary and reserve forces amount to a further 718,000 but, of these, 700,000 are in the Youth Labor Army, Civil Defense Force or Territorial Militia. Cuban military personnel also serve overseas and in 1983/4 it was estimated that there were 26,000 Cuban military 'advisers' in Angola, 750 in Congo-Brazzaville, 750 in Mozambique, 300 in South Yemen, 1000 in Nicaragua and a further 500 in other African states such as Algeria, Guinea-Bissau, Libya, the Malagasy Republic, Tanzania and Zambia. In 1983/4 some 12,500 were also in Ethiopia but in the course of 1984 this has declined to only an estimated 2000. Other Cuban advisers, who are technicians or other non-military personnel, have also been reported in these aforementioned states as well as Afghanistan, Iraq, and Saõ Tome. In Nicaragua, for example, the total Cuban presence may amount to over 6000 men.

Nicaragua

Nicaragua itself has armed forces totalling 48,800 with 47,000 serving in the army, 1500 in the air force and 300 in the navy. The navy has 14 patrol craft and, so far, the air force is believed to operate only 10 combat aircraft, all trainers converted to an anti-guerrilla role. There have been persistent reports since 1982 that Nicaraguans have received pilot training in Eastern Bloc states such as Bulgaria and that MiG combat aircraft are about to be or have been supplied. There is no firm evidence for such reports, the most recent being in November 1984. Like the Grenadian situation, the United States is also suspicious of the construction of a 20,000ft runway in Nicaragua. The Nicaraguan army, backed by up to 25,000 reserves at any one time, is organized into three armored battalions, 10 infantry battalions, an engineer battalion, and a field artillery group. Its main armament consists of 48 tanks – three ageing US M4 Shermans and 45 T-54/55s.

Guerrilla groups remain active in many countries of Latin America but few are particularly significant except the 6000 guerrillas of FMLN in El Salvador, who are estimated to control some 30 percent of the country. Another possible future threat is the group known as *Sendero Luminoso* ('Shining Path'), a messianic Maoist revolutionary group active in Peru since 1980. It has some 3000 adherents and has apparently rediscovered the virtues of rural insurgency discredited by the numerous failures of the 1960s.

Prospects for the future

Clearly, then, Central America remains a crisis area at the present time but there are some signs of accommodation. In El Salvador, for example, the United States has invested considerable expectations in the Christian Democrat, José Napoleón Duarte. A former presidential candidate imprisoned and tortured by the military, Duarte returned from exile and joined the government of officers who had overthrown Romero. He was president for some 16 months but was judged to have lost a decidedly inconclusive election in March 1982. Nevertheless, some progress was made and Duarte himself defeated the candidate of the extreme right, Roberto D'Aubuisson, in the elections of June 1984. Negotiations began with the FMLN in October 1984 although this has not resulted in any cessation of a war that has cost an estimated 45,000 lives since 1979. Nevertheless, the FMLN itself appears to be ready for some continuing dialogue and has claimed, most recently, that the guerrillas would be prepared to offer guarantees to the United States in terms of El Salvador not becoming a foreign base or a surrogate of the Soviet Union.

In Nicaragua, the Sandinistas have accepted a compromise peace plan put forward by the so-called CONTADORA group of Mexico, Venezuela, Colombia and Panama, Contadora being the Panamanian island where the delegations first met in February 1983. This involves

the withdrawal of both United States and other foreign advisers from Nicaragua and its neighbors. The United States has not so far shown any interest in the proposals and the reports of the imminent arrival of MiG combat aircraft in Nicaragua in November 1984 and the counter claims that the United States was poised to invade Nicaragua have not aided negotiation.

It is apparent that the Soviet Union has been notably cautious in its dealings with Central and South America. The Soviets would be loath to repeat the economic commitment they have been forced to extend to Cuba, which is now estimated at $3 billion a year or approximately one quarter of Cuba's GNP. The Soviets still lack real 'reach' in the region and have largely become involved only through invitations to do so. They would probably support revolution if no other course presented itself but would not be prepared to do so openly. Cuba, too, appears to favor a negotiated settlement in Nicaragua and Latin America generally no longer appears to view Castro with the same alarm that it did in the 1960s. Cuba received kudos from offering troops to Argentina during the Falklands conflict in 1982 and, significantly, Argentina withdrew its assistance to the government of El Salvador in protest at United States support for Britain. There is also little doubt that the members of the OAS were perturbed by the stance adopted by the Reagan administration at that time. In short, the Soviets and the Cubans need do little but capitalize upon the mistakes made by the United States in its own 'backyard.'

Sadly, recent and past history is littered with examples of miscalculations by the United States in the region. It is arguable that Cuba in 1959 or Grenada and Nicaragua in 1979 might have developed rather differently given a more pragmatic reaction in the United States. Realism coupled with restraint is required in addition to continued economic assistance and encouragement of genuine reform among the often unsavory governments of Latin America. In this regard the future of Grenada, where Eric Gairy initially seemed poised for a comeback, may prove a test case of United States interest in promoting local democracy. Given greater receptiveness to the mood of Central and South American states on the part of the United States, there is no reason to suppose that the threat at present posed by communism will not be contained. Indeed, it might well be concluded that if communism should arrive on the very doorstep of the United States by way of Central America and Mexico, it would, in all probability, be a result of United States policy rather than Soviet or Cuban design.

Above: Castro's brother, Raul (center), with other eastern bloc observers at Warsaw Pact maneuvers in September 1981 in his capacity as Cuba's Armed Forces Minister.

CONCLUSION

Above: The leaders of the Politburo taking the salute at the annual military parade in Red Square in Moscow to celebrate the anniversary of the Revolution. The parade of 7 November 1983 is shown.

The vast land mass that stretches from the river Elbe in central Europe to the shores of the Pacific is dominated by communism, from the Soviet Union and its satellites of eastern Europe to the People's Republic of China and the communist states of Southeast Asia headed by Vietnam. Of these, China is by far the most populous state in the world with an estimated population in excess of one billion while the Soviet Union embraces close to one sixth of the entire land surface of the globe within its frontiers. Beyond this heavy concentration of communism of differing kinds, there are well over 60 other communist parties in other parts of the world, of which over 40 are clandestine or illegal or both. By comparison the number of genuinely democratic governments in the world is few although including, in the shape of the United States, the other great global nuclear superpower besides the Soviet Union. Communism is no longer perceived to be the monolithic structure it appeared immediately after World War II, when communism first emerged as a major factor in global affairs, but it still represents a formidable challenge to the free world.

That communism is not monolithic has become increasingly obvious, the Sino-Soviet split being one of the most important developments in the postwar world. Others, too, have gone their own way such as Josip Broz Tito in Yugoslavia, Enver Hoxha in Albania and Kim Il Sung in North Korea. The conference of communist parties held at Budapest in June 1969 – an attempt to show a united front after the Soviet invasion of Czechoslovakia in the preceding year – marked the clear emergence of what has become known as 'polycentrism.' Five of the world's then 14 ruling communist parties declined to attend at all – China, North Korea, North Vietnam, Albania and Yugoslavia – while 13 non-ruling communist parties also stayed away. Of the 75 that did attend, only 61 signed the final communiqué without reservation. Many of the world's non-ruling communist parties are in no position to achieve results even if they acknowledge the leadership of Moscow. Clearly, too, not all ruling communist parties represent quite the same degree of threat to the free world as others.

The People's Republic of China is a case in point. China lacks the capability of posing a significant military threat to either of the superpowers. The Chinese may well have a limited second strike nuclear capability – sometimes referred to as a 'bee-sting' deterrent – but this can only represent a long-term challenge to the Soviet Union and the United States. Of more significance is the fact that the Chinese have chosen for themselves an essentially didactic role in world affairs. They do not seek to force others to accept their view of the future, unless China's territorial integrity is deemed to be under threat. Following what has been described as a variant of Stalin's 'socialism in one country,' the Chinese are content to offer an example of a revolutionary model to others. China is thus a 'revolutionary base area' engaged in a 'massive educational exercise' for which the immediate prospects of success appear slight. But China is far more concerned with her own process of modernization which may ultimately enhance the capabilities of a state which is already a nuclear power in its own right, one of only five permanent members of the United Nations Security Council and undoubtedly a great power.

That potential for the future is what concerns the Soviet Union, which views with alarm the loose coalition which has developed between the United States, western Europe, Japan and China. To a large extent, of course, Soviet foreign policy is governed by a sense of insecurity stemming both from Russia's history and from a general developmental inferiority complex when the Soviet Union is compared to the West. In terms of the latter factor, it can be noted that the idea that the West does not deserve its advanced standard of living is shared by those adhering to the Russian Orthodox Church just as much by those adhering to Marxist-Leninism since Moscow is regarded by the Orthdox Church as the 'third Rome' to which the center of world Christianity devolved following the fall of Constantinople. In terms of Russia's past, Soviet perceptions owe much to the consciousness of the insecurity of land frontiers and the many past invasions of Mother Russia. The majority of such invasions have come from the west, hence the traditional interest in securing eastern Europe as a buffer zone, but the last successful invasion came from the east in the form of Genghis Khan. Numbers have always counted in Russian history and that merely adds to the alarm with which the Soviets view the Chinese 'wall of humanity.' It does not help that the proportion of European Slavic Russians within the Soviet Union is falling by comparison to the faster birth rate of non-Slavs. Superficially, even geography further conspires to confirm the Soviet concept of encirclement by the rest of the world since Soviet ideas derive from the very different perspective afforded them (compared to the Mercator projection with which westerners are most familiar) in Soviet maps of the world. The ideological and

Above: Afghan soldiers loyal to the communist government in a Kabul street. As is often the case in a guerrilla war the government forces control the towns but have many problems in the countryside.

historical perspectives are yet further sharpened by the belief that 'Western imperialists' are inherently hostile to socialism and continually striving to 'reverse the course of history.'

The sense of insecurity breeds a natural inclination to overcompensate and over-insure on security matters while, additionally, 'ideological symbolism' also affects Soviet thinking in terms of a perceived requirement for superior military capability to all other states. The Soviets, beginning from a position in 1945 of nuclear inferiority to the United States, single-mindedly pursued strategic nuclear parity, which they had achieved by the mid 1970s. By the early 1980s the Soviets had attained what some Americans have categorized as strategic equivalence and may have gained strategic superiority in some areas of the nuclear balance. Through its large land mass and concomitant proximity to many strategically important areas, the Soviet Union also has the ability to project its military power over almost half the globe without the assistance of overseas base facilities. Furthermore, the development of the Soviet Navy since the 1960s has extended that ability to project power and influence.

Yet, at the same time, the Soviet Union is a flawed giant. The vastness of the Soviet Union itself is a source of weakness, difficulties arising in terms of communication, combating regionalism and ensuring centralization of control from Moscow. The last of the European empires, the Soviet Union's frontiers embrace many races and some 122 different languages are spoken within them. Other domestic problems have also been encountered. Much of the protein in the diet of the Soviet population is derived from fish that can only be caught off other peoples' coasts. Soviet agriculture has been grossly inefficient, some six million tonnes of grain being imported annually from the United States in the 1970s until President Carter's embargo in the wake of the Soviet invasion of Afghanistan in December 1979. The Soviet Union has also been compelled to import substantial quantities of foreign technology to reverse her declining industrial productivity. To pay for such imports, the Soviets must export increasing amounts of their own raw materials and are heavily in debt to the West as are many of the eastern European satellites, the possible political instability of which under continuing social and economic pressures is another source of concern. To maintain such exports of raw materials, the Soviets must tap more inaccessible resources and reserves which, in turn, can only be achieved with more western technological expertise. Economic growth has continued to decline and it is believed that, in the 1980s, the Soviet Union will begin to suffer an energy crisis, having previously been a net exporter of energy.

Such growing problems may conceivably compel the Soviet Union to devote more of its resources to solving more pressing domestic difficulties in the future. It was argued that Yuri Andropov, who succeeded Leonid Brezhnev as

Below: Babrak Karmal waves to the crowd during a parade in Kabul in April 1980 held to celebrate the Afghanistan revolution.

Soviet leader in November 1982, was primarily concerned with rooting out inefficiency and corruption within the Soviet system. Andropov, of course, died in February 1984 whereupon it would appear that his successor, the 72 year old Konstantin Chernenko, who died himself in March 1985, was less interested in any anti-corruption drives. It is often suggested that younger members of the ruling Politburo led by Mikhail Gorbachev may again turn to domestic issues now that power has come their way. Certainly, the few remaining members of the old gerontocracy must soon lose all influence, seven of the surviving 12 full members of the Politburo after Andropov's death being aged over 70 years of age, with an average age of over 67 for all 12. However, it has also been suggested that a younger leadership would be no less bound by the historical obligation to promote the spread of communism and that it might conceivably be even more willing than the relatively cautious leadership of recent years to seize opportunities presented by the strength and confidence of the Soviet armed forces, in order to achieve a measurable improvement of the Soviet position as insurance against an uncertain future. A totalitarian state such as the Soviet Union has fewer internal restraints on its actions than western democracies and there is little doubt that Soviet military capabilities have significantly improved, as may be gauged from a comparison of performance in Czechoslovakia in 1968 with that in Ethiopia in 1977 or Afghanistan in 1979.

What can be predicted is that the Soviet Union, whatever the complexion of its leadership, will continue to make the most of any opportunities presented. Such opportunities have been seized in the past irrespective of any ideological considerations in terms of local conditions. In the process the Soviets have not always succeeded in transcending local nationalism or local rivalries and there have been failures, particularly in the Third World. This is not necessarily important since many parts of the Third World are of only incidental significance to the Soviet Union. However, in the Middle East where the Soviets do believe that they have interests to safeguard, militant Islamic fundamentalism has gelled uneasily with Marxist-Leninism. Opportunism has also brought long-term military or financial commitments as in Cuba, Ethiopia and Vietnam although it should also be pointed out that Afghanistan, even by the highest estimates, detains only five percent of Soviet divisional troops. Nor have the clients of the Soviet Union always proved reliable, only the People's Democratic Republic of the Yemen appearing mostly satisfactory in this respect. As it happens, South Yemen is also that client best placed to cause damage in an area of great strategic significance for both the West and the Soviet Union. Outside the eastern bloc in Europe, the Soviets have concluded treaties of friendship and cooperation with North Korea (1961), Mongolia (1966), India (1971), Iraq (1972), Angola (1976), Mozambique (1977),

Below: Libya's Colonel Qaddafi and Nicaragua's Daniel Ortega attending Libyan maneuvers in 1981.

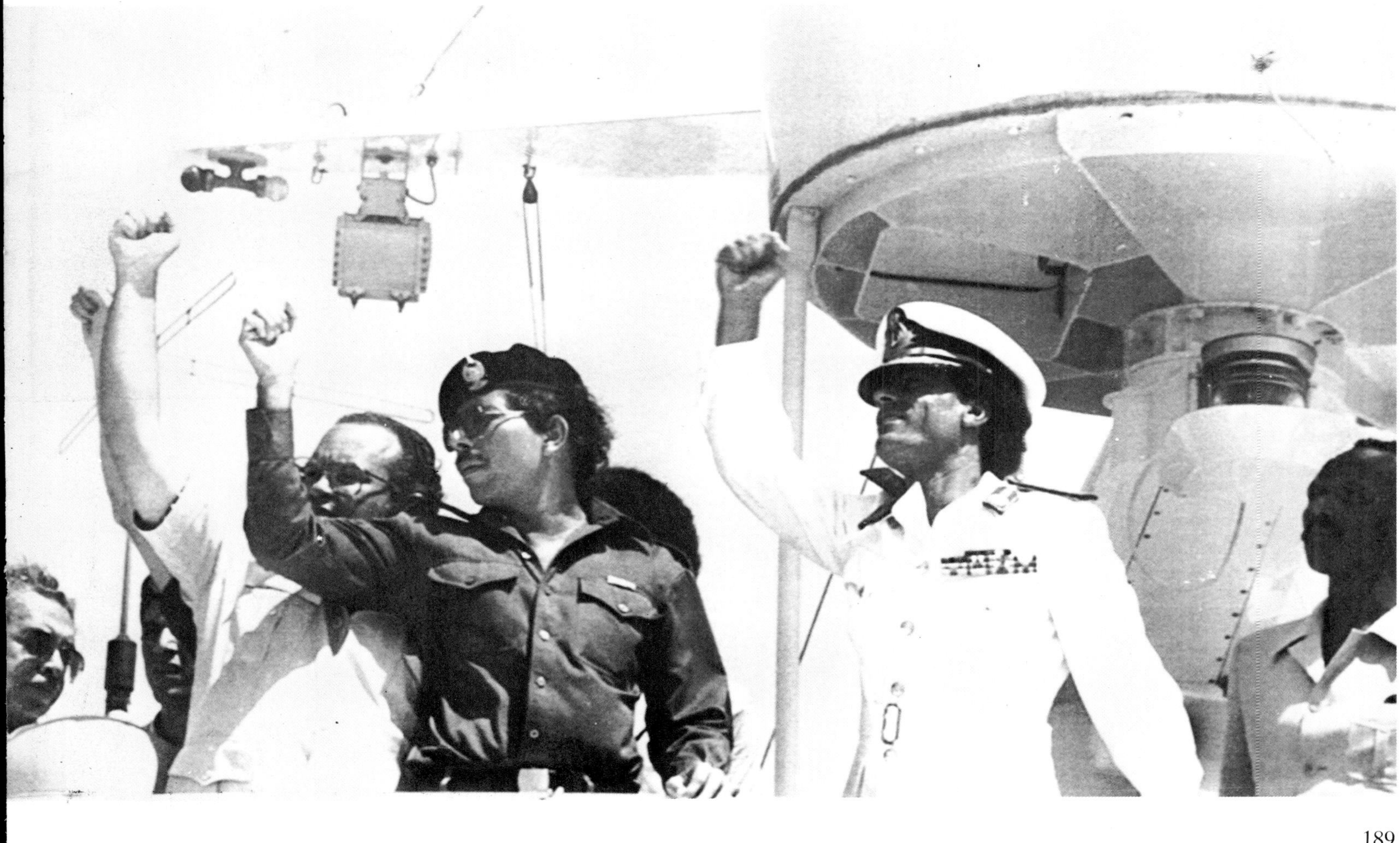

Above: Soviet troops in BMP armored fighting vehicles. The BMP is armed with Sagger anti-tank missiles and a 23mm gun. Eight soldiers are normally carried in addition to the three-man crew.

Ethiopia (1978), Afghanistan (1978), Vietnam (1978), South Yemen (1979) and Syria (1980). There is no formal treaty with Cuba and the Soviets actually turned down a treaty offered by the erratic Colonel Qaddafi of Libya in 1981.

Nevertheless, some of the opportunities and risks taken have paid off handsomely and have served the purpose of the Soviets in undermining western influence, offsetting Chinese influence and fulfilling strategic requirements. In many cases such results have been achieved at low cost and low risk through the failure of the West to respond adequately or at all. Open confrontation with the West is something the Soviets wish to avoid and they have always been interested in the demilitarization and 'finlandization' of their immediate neighbors. They have thus offered nuclear free zones to Scandinavia, a 'system of collective security' in the Indian Ocean and have unsuccessfully courted Japan. But, if other means are not available, the Soviets will resort to military power. This may be covert support for terrorism, action through surrogates or proxies or even direct military action as in 1953, 1956, 1968 and 1979. Thus, although a high priority is given to avoiding war, the Soviets are intent that any war waged should be won, even a nuclear exchange.

Opportunism on the part of the Soviet Union is not necessarily to be equated with adventurism or a conspiracy to destroy the free world but that opportunism is underpinned by the ultimate ideological goals of Marxist-Leninism. Senator Henry 'Scoop' Jackson once likened the Soviet Union to a burglar trying the doors along a hotel corridor. If the door was locked then the Soviet Union would go away but, if it was open, the Soviets would go in. The lesson for the West is that too many doors have been left unlocked in the past through western neglect or indecision. The Soviet Union respects only strength – it despises weakness.

Index

Acknowledgments

The publishers would like to thank Adrian Hodgkins who designed this book and Ron Watson who compiled the index. The following agencies and individuals kindly supplied the illustrations.

ADN: page 135 (bottom).
AFP: pages 108-9.
Associated Press: pages 78-9, 79 (bottom), 80, 83, 84, 86 (both), 87, 88-9, 89, 92, 93 (bottom), 97 (top), 100-1, 101 (bottom), 102, 104 (top left), 106, 107 (top), 112 (both), 113 (top), 116 (inset), 120, 121, 124, 124-5, 129, 132 (top), 138-9, 145, 147, 152 (both), 155 (bottom), 158 (bottom), 164 (bottom), 175, 177, 180, 181 (bottom).
Bison Picture Library: pages 6, 7 (both), 11, 12, 13, 14, 14-15, 16 (both), 17 (bottom), 18-19 (all three), 20, 26 (both), 27 (both), 28, 29 (top right & bottom right), 30-1 (all four), 33 (both), 34-5 (main pic), 34 (inset right), 35 (inset), 36 (both), 37, 40-1 (both), 42 (middle & bottom), 43, 48 (both), 50-1 (both), 54, 54-5, 56 (top), 57 (top), 58-9 (all four), 60 (top), 62-3 (all five), 64 (both), 66-7, 70-1 (both), 73, 74-5 (both), 90, 91, 97 (bottom), 114 (both), 115, 118-19, 122, 123, 126 (both), 127 (bottom), 130, 149 (top), 166, 179 (bottom), 182, 183 (both), 190, back jacket.
Bison/DFS Pictorial: page 93 (top).
China Photo Service: page 69.
Comlit Press/Defence Magazine: page 79 (top).
SPC Dalziel: pages 47 (both), 55 (top).
Defence Magazine: pages 65, 178 (top left), 189.
ECPArmées/MARS, Lincs: page 95 (top).
Archiv Gerstenberg: pages 8, 9 (top).
Helmoed-Römer Heitman: page 154.
M Hooks: pages 38-9.
Impact Photos/Alain le Garsmeur: page 179 (top).
Imperial War Museum: page 77 (top).
Mansell Collection: pages 9 (bottom), 17 (top).
Mediaco: page 162 (top).
MOD, London: pages 2-3.
Peter Newark's Western Americana: page 68.
Novosti Press Agency: front jacket, pages 10, 22-3, 24, 25 (bottom right), 29 (top left), 53, 57 (bottom), 61 (top), 72, 107 (bottom), 125 (bottom), 127 (top), 131, 165, 186, 187, 188.
Paratus: pages 150/1.
The Photo Source/Central Press: pages 82-3, 88 (top left), 142 (top), 144, 172-3 (top).
The Photo Source/Keystone: pages 1, 56 (bottom), 76 (both), 81, 85, 88 (bottom), 101 (top), 104 (bottom), 104-5, 105 (right), 108 (left), 109 (bottom), 110-11 (both), 113 (bottom), 116 (main picture), 128, 132 (bottom), 136 (both), 137, 141, 142 (bottom), 143 (both), 146, 148, 151 (all three), 153, 155 (top), 156-7 (both), 158 (top), 159 (both), 160, 161, 162-3, 167, 168 (both), 169, 170 (top left), 173 (bottom), 176 (both), 178 (bottom).
Steenkamp Collection: pages 149 (bottom), 164 (top).
Syrian Government Press Office: page 133.
TASS/Defence Magazine: pages 4-5, 25 (top left, top right & bottom left), 28 (top left & top right), 185.
US Air Force: pages 170-1 (both), 171 (top right).
US Army: pages 95 (bottom), 98-9 (all three), 178 (middle left & right).
US Army/Defence Magazine: pages 34 (inset left), 96.
US Department of Defense: pages 39 (inset), 77 (bottom), 134, 135 (top).
US Department of Defense/Defence Magazine: page 42 (top).
US Department of Defense/MARS, Lincs: page 181 (top).
WWP: pages 44-5 (all three), 46, 52, 60-1.